Remote Sensing and Actuation Using Unmanned Vehicles

IEEE Press Series on Systems Science and Engineering

A complete list of the titles in this series appears at the end of this volume.

Remote Sensing and Actuation Using Unmanned Vehicles

Haiyang Chao

YangQuan Chen

IEEE PRESS

A JOHN WILEY & SONS, INC., PUBLICATION

Library of Congress Cataloging-in-Publication Data:

Chao, Haiyang.
  Remote sensing and actuation using unmanned vehicles/Haiyang
Chao, Yangquan Chen.
      pages cm
    Includes bibliographical references and index.
    ISBN 978-1-118-12276-1 (hardback)
  1. Geomorphology–Remote sensing.   2. Environmental monitoring–Remote
sensing.   3. Vehicles, Remotely piloted.   I. Chen, Yangquan, 1966-   II. Title.
    GB400.42.R4C46 2012
    621.36'78–dc23
                                                                    2012006660

To my parents and my brother
Haiyang Chao

To my family
YangQuan Chen

Contents in Brief

Contents

List of Figures

List of Tables

Foreword

This monograph is very timely and in many ways prophetic. In the United States, lawmakers just passed a long awaited bill setting requirements to integrate unmanned aerial systems (UAS) in the national airspace (NAS). Such requirements would set deadlines for full-scale integration in fall 2015. News of UAS also fills headlines as both their mission roles and numbers increase. It is estimated that over one-third of military aircraft in the United States are unmanned. As these numbers continue to grow, coupled with the opening of the NAS, challenges in technology, policy, and even ethics will remain in the forefront for years to come. As such, this monograph provides a much needed resource to root the community; academics, industrial practitioners, policymakers, and business developers stand to benefit on the foundations presented in this book.

Aviation giants all have divisions that are aggressively pursuing UAS programs. The conviction is that aerospace leadership into the 21st century will be determined by those who successfully commercialize unmanned aircraft. Such mindsets are thus pushing the boundaries of UAS roles with creative applications to meet untapped needs such as agricultural crop handling, wildlife monitoring, road traffic reporting, meteorological pattern detection, and forest wildfire mitigation. Realization of such applications is still underpinned by open research challenges. This monograph provides insight into both the engineering fundamentals and best practices that have resulted in years of field-proven work. The authors Haiyang Chao and YangQuan Chen are pioneers in UAS research and development. They also led award-winning teams in international UAS competitions and have logged years of flight time. Readers are indeed fortunate to benefit from their experiences as vetted in this book.

This book is striking in its clarity and comprehensiveness. Moreover, the book is underscored by examination and examples revolving around a low-cost UAS. From aerodynamic modeling fundamentals to actual construction and flight-testing, readers can implement their own real-world UAS. Such a UAS provides a focal point and test bed to apply topics like flight control, wide-area coverage, multi-UAV formations and aerial image processing.

Here to now, UAS design, build and fly has been mostly ad hoc. This book stands to be the definitive guide to analytically design and systematically engineer UAS. As such, this book will be a must for every individual including UAS designers and users.

This is a very timely book and will help shape an exciting future as UAS become part of our everyday experience.

Paul Y. Oh, *ASME Fellow*

Founding Chair of IEEE RAS TC on ARUAV
Associate Professor & Department Head
Mechanical Engineering & Mechanics Department
Drexel University, Philadelphia, PA, USA

Preface

Unmanned vehicles, including unmanned aerial vehicles (UAVs) and unmanned ground vehicles (UGVs), have been increasingly used to free human beings from dangerous, dull, and dirty jobs. Unmanned vehicles can serve as remote sensors for surveillance applications, or actuators for control purposes, or both. Typical remote sensing tasks include the monitoring of air quality, forest fire, and nuclear leaks, which all require measurements over a large scale (typically tens of square miles or even bigger). Example actuation tasks include fire extinguishing, and neutralizing control of chemical spills. One of the key challenges in unmanned vehicle developments is the design of autonomous capability or remote control capability. More importantly, unmanned vehicles need to combine navigation, communication, and computation capabilities with mission-specific payload to provide useful information. For instance, each aerial image requires synchronized position and orientation information for image interpretation purposes. Another key challenge using unmanned vehicles is how to deploy single or groups of vehicles optimally for different sensing or actuation tasks. Optimal paths are needed to extend the sensing range of unmanned vehicles, or to minimize the time taken for control missions.

This monograph focuses on how to design and employ unmanned vehicles for remote sensing and distributed control purposes in the current information-rich world. The target scenarios are environmental or agricultural applications such as river/reservoir surveillance, wind profiling measurement, monitoring/control of chemical leaks, and so on. This monograph comprises two parts. The first part is on the design and deploy of UAVs for remote sensing applications with a focus on real flight experiments. AggieAir, a small and low-cost unmanned aircraft system, is designed based on the remote sensing requirements from environmental monitoring missions. A summary of low-cost IMU hardware and attitude estimation software is detailed for small UAVs. The advanced lateral flight controller design problem is further introduced. Then the single-UAV-based remote sensing problem is focused with flight test results. The latter part of this monograph is on ground test and simulations to use networked unmanned vehicles in future sensing and control missions such as formation control, wind profile monitoring, and neutralizing control. Given the measurements from unmanned vehicles, the actuation algorithms are needed for missions such as the diffusion control. A consensus-based central Voronoi tessellation (CVT) algorithm is proposed for better control of the diffusion process. Finally, the monograph conclusion and some new research suggestions are presented. A Web site has been set up with extra videos, figures, and source codes (http://sites.google.com/site/haiyangchao/book_rsauv).

This book can benefit engineers or developers who want to develop their own low-cost UAV platforms for remote sensing, environmental monitoring, aerial image processing, wireless communication, flight control, cooperative control, or general robotics researches. More importantly, this book provides the authors' approach on using unmanned vehicles for formation control and actuation missions. Such missions will be very important especially in scenarios such as nuclear leaks. The authors hope that this book can help increase the uses of unmanned vehicles in both remote sensing and actuation applications.

HAIYANG CHAO,
YANGQUAN CHEN

January, 2012

Acknowledgments

The authors would like to thank other contributors of the AggieAir UAS development including Austin M. Jensen, Yiding Han, Long Di, Calvin Coopmans, and Daniel Morgan. Part of the results in Chapter 5 was collected with AggieAir platform by Austin M. Jensen from Utah Water Research Laboratory. The fractional flight controller is codeveloped by the authors together with Dr. Ying Luo and Long Di. The authors also want to thank other past OSAM UAV team members: Marc Baumann, Dr. Yongcan Cao, Dr. Hu Sheng, Aaron Avery, Mitch Humphries, Chris Hall, Aaron Quitberg, Norman Wildmann, and other MAS-net team members: William K. Bourgeous, Nathan Sorensen, Dr. Zhen Song.

Haiyang Chao would like to thank Dr. Wei Ren for joint work in the experimental validations on the MAS-net platform; Dr. Donn Cripps, Dr. Todd Moon, Dr. Vladimir Kulyukin, Dr. Jiming Jin and Dr. Hui Chen for their help in his Ph.D. dissertation work. Thanks also go to other CSOIS members for their help including Dr. Hyo-Sung Ahn, Dr. Christophe Tricaud, Rongtao Sun, Yashordhan Tarte, Tripti Bhaskaran, Varsha Bhambhani, Jessi Lianghan N.G., Shayok Mukhopadhyay, Dr. Yan Li, Dr. Bin Wang, Dr. Wei Sun, Dr. Yongshun Jin, Dr. Hongguang Sun. The authors want to especially thank MaryLee Anderson, Brandon Stark, and Jinlu Han for proof reading, Dr. Dongbing Gu for providing very useful reviewer comments, Mary Hatcher and Taisuke Soda from John Wiley & Sons for helping in publishing this book.

Last but not the least, the authors want to thank developers from the open-source society including Curtis Olson, Pascal Briset, Gautier Hattenberger, Anton Kochevar, Antoine Drouin, Felix Ruess, William Premerlani, Paul Bizard, Dr. JungSoon Jang, and all the Paparazzi UAV developers for sharing their open-source projects. The authors benefited a lot through reading their source codes and discussing details with them.

The work presented in this book is partly supported by the Utah Water Research Laboratory (UWRL) MLF Seed Grant (2006–2011) and the Vice President for Research Fellowship (2005–2006) from Utah State University. The authors are thankful to Professor Mac McKee for his original research vision on UAV application in water science and engineering and Professor Raymond L. Cartee for providing the USU farm at Cache Junction as the flight test field, and Professor H. Scott Hinton, the Dean of College of Engineering, for travel support to Maryland for the AUVSI Student UAS competitions in summer 2008 and 2009. Haiyang Chao would like to thank the Graduate Student Senate of Utah State University for the Enhancement Award, and the NSF Institute for Pure and Applied Mathematics (IPAM) at UCLA for the travel support to participate in the 1 week-long workshop on "Mathematical Challenges on Sensor Networks."

Acronyms

AGL	above ground level
ARX	auto-regressive exogenous
AUVSI	Association for Unmanned Vehicle Systems International
BFCS	body-fixed coordinate system
CCD	charge-coupled device
CG	center of gravity
COTS	commercial off-the-shelf
CPS	cyber-physical systems
CSOIS	Center for Self-Organizing and Intelligent Systems
CVT	centroidal Voronoi tessellations
DC	digital camera
DCM	direction cosine matrix
DOF	degree of freedom
DPS	distributed parameter system
DV	digital video camera
ECEF	Earth-centered Earth-fixed
ENAC	Ecole Nationale de l'Aviation Civile
EKF	extended Kalman filter
FOC	fractional order control
FOPI	fractional order proportional integral
FOPTD	first order plus time delay
FOV	field of view
GA	genetic algorithm
GCS	ground control station
GF	Ghost Finger
GF-DC	Ghost Finger digital camera
GF-DV	Ghost Finger digital video camera
GIS	geographic information system
GPS	global position system
gRAID	geospatial real-time aerial image display
ID	identification
IGS	inertial guidance system
IMU	inertial measurement unit
INS	inertial navigation system
IOPI	integer order proportional integral
IR	infrared
LQG	linear quadratic Gaussian

LIDAR	light detection and ranging
LLH	latitude longitude height
LS	least squares
LTI	linear time-invariant
LTV	linear time-varying
MASnet	mobile actuator and sensor network
MEMS	microelectromechanical systems
MIMO	multiple inputs multiple outputs
MZN	modified Ziegler–Nichols
NIR	near infrared
NN	neural network
ODE	ordinary differential equation
OSAM-UAV	open-source autonomous multiple unmanned aerial vehicle
PDE	partial differential equation
PI	proportional-integral
PID	proportional-integral-derivative
PPRZ	Paparazzi open-source autopilot project
PRBS	pseudo random binary signal
PTP	picture transfer protocol
RC	remote controlled
RGB	red–green–blue
RPV	remote piloted vehicle
SD	secure digital
SISO	single input single output
SO3	special orthogonal group 3
TWOG	Tiny without GPS
UART	universal asynchronous receiver/transmitter
UAS	unmanned aircraft system
UAV	unmanned air/aerial vehicle
UGV	unmanned ground vehicle
USB	universal serial bus
USU	Utah State University
UTM	universal transverse mercator
UUV	unmanned underwater vehicle
UWRL	Utah Water Research Lab
VC	video camera
WVU	West Virginia University
XML	extensible markup language

Chapter 1

Introduction

1.1 MONOGRAPH ROADMAP

This monograph focuses on how to design and employ unmanned systems for remote sensing and distributed control purposes in the current information-rich world. The target scenarios include river/reservoir surveillance, wind profiling measurement, distributed control of chemical leaks, and the like, which are all closely related to the physical environment. Nowadays, threats of global warming and climate change demand accurate and low-cost techniques for a better modeling and control of the environmental physical processes. Unmanned systems could serve as mobile or stationary sensors and actuators. They could save human beings from dangerous, tedious, and repetitive outdoor work, whether it is deep in the ocean or high up in the sky. With the modern wireless communication technologies, unmanned vehicles could even work in groups for some challenging missions such as forest fire monitoring, ocean sampling, and so on. However, unmanned systems still require physics-coupled algorithms to accomplish such tasks mostly in the outdoor unstructured environments. Questions such as what to measure, when to measure, where to measure, and how to control all need to be properly addressed. This monograph presents our approach about how to build and employ unmanned vehicles (ground, air, or combined) to solve the problem of distributed sensing and distributed control of agricultural/environmental systems.

1.1.1 Sensing and Control in the Information-Rich World

Advances in electronics technologies such as embedded systems, microelectromechanical systems, and reliable wireless networks make it possible to deploy low-cost sensors and actuators in large amounts in a large-scale system. This poses a problem for control scientists and engineers on how to deploy and employ those vast amount of networked sensors/actuators optimally. The sensors and actuators can be static or mobile, single or multiple, isolated or networked, all depending on the application scenario. The options for sensor and actuator types are shown in Fig. 1.1. For example, both the temperature probe (point-wise sensing) and the thermal camera

Remote Sensing and Actuation Using Unmanned Vehicles, First Edition. Haiyang Chao and YangQuan Chen.
© 2012 by The Institute of Electrical and Electronics Engineers, Inc.
Published 2012 by John Wiley & Sons, Inc.

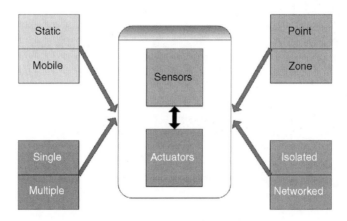

Figure 1.1   Sensors and actuators in an information-rich world.

(zone sensing) could be used to measure the temperature of the crop canopy in a given field of interest. But which one to use? Proper sensing techniques are essential for the high-precision farming that can support the sensing of a large-scale system with an acceptable cost. Thermal aerial images are better for this mission. On the other hand, there are also coarse agricultural applications, which only need the temperature probe due to the cost limits. Another typical example is to use unmanned vehicles to monitor the forest fires. It is intuitive to use multiple unmanned aerial vehicles (UAVs), since they could provide more real-time information. However, there are questions regarding what information to share among UAVs and how often to share.

Unmanned vehicles can add the mobility to the sensors and actuators, which is especially beneficial for most outdoor environment monitoring applications. Different kinds of sensors and actuators could be installed on the unmanned vehicles based on specific application scenarios, as shown in Fig. 1.2. For instance, contact sensors can be installed on unmanned underwater vehicles (UUVs) to make accurate measurements of the temperature and humidity of the sea current. Cameras or radars can be mounted on UAVs for a more complete view of a farm or a reservoir. Chemical sprayers could be installed on unmanned ground vehicles (UGVs) for neutralizing gas leaks or extinguishing fires.

In this monograph, the unmanned system is defined as the unmanned vehicle together with onboard payload sensors or actuators. The fundamental functions of a typical unmanned systems include the mobility, computation, decision making, communication, and sensing/actuation, as shown in Fig. 1.3. Most unmanned systems have a powerful embedded processor to coordinate all the functions and make decisions based on information collected from its own or shared from other neighboring vehicles. With the communication subsystems, groups comprising of heterogeneous unmanned systems can now be designed to cooperate with each other to maximize their capabilities and the team's collective performance.

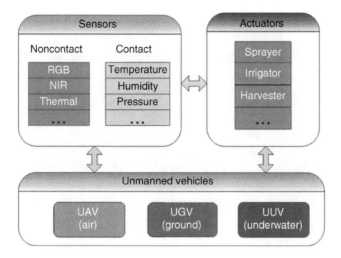

Figure 1.2   Unmanned vehicles as mobile sensors/actuators.

1.1.2   Typical Civilian Application Scenarios

This monograph focuses mostly on the monitoring and control of environmental or agricultural systems or processes, which are of course closely related to human beings. Such systems could be categorized into two groups: fast-evolving ones such as chemical spill, gas leak, or forest fire and slow-evolving ones including heat transfer, moisture changing, wind profiling, and the like. The objective of monitoring these kinds of systems is to characterize how one or several physical entities evolve with both time and space. One typical example is an agricultural farm, as shown in Fig. 1.4. Water managers are interested in knowing how the soil moisture evolves with time in a farm to minimize the water consumption for irrigations. However, the evolution of

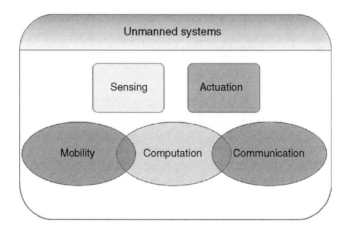

Figure 1.3   System structures of unmanned vehicles.

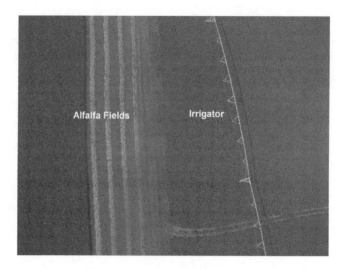

Figure 1.4 Typical agricultural field (Cache Junction, UT). (*See insert for color representation of this figure.*)

soil moisture is affected by many other factors such as water flows, weather conditions (e.g., wind), and vegetation types, which all require measurements over a large scale (typically tens of square miles or even bigger). For such missions, ground probe stations are expensive to build and can only provide sensor data with very limited range. Satellite images can cover a large area, but have a low spatial resolution and a slow temporal update rate. Small UAVs cost less money but can provide more accurate information from low altitudes with less interference from clouds. In addition, small UAVs combined with ground and orbital sensors can form a multiscale remote sensing system, shown in Fig. 1.5.

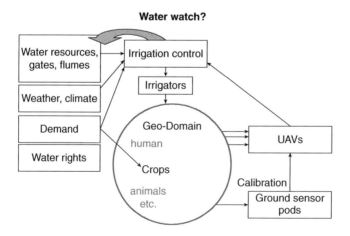

Figure 1.5 Water watch concept.

Figure 1.6   Fog evolution (taken in Yellowstone National Park). (*See insert for color representation of this figure.*)

Other typical civilian applications of unmanned systems include:

- *Forest Fire Monitoring and Containment Control*: The monitoring, prediction, and containment control of forest fires could greatly reduce the potential property damages. Unmanned systems have obvious advantages over manned vehicles because human operators are not required onboard.

- *Fog Evolution or Chemical Leaking Monitoring and Control*: The evolution of hazardous fogs under emergency conditions can cost human lives without accurate and real-time measurements from unmanned systems. Example harmless fog evolutions are shown in Fig. 1.6.

- *Wind Field Measurement*: The wind direction and wind speed could have a significant impact on the diffusion of heat, water, or wind powers. However, the wind field is hard to measure because of its high variation, both temporally and spatially. Unmanned vehicles can be easily sent into the air for accurate 3D measurements.

- *Canopy Moisture Measurement and Irrigation Control*: The moisture on the vegetation canopy represents how much water could be absorbed by the plants. This information can be used for accurate irrigation control. The large scale of most agriculture fields requires cheap sensing techniques.

1.1.3 Challenges in Sensing and Control Using Unmanned Vehicles

The problem of monitoring an environmental field can be defined as below. Let $\Omega \subset R^3$ be a polytope including the interior, which can be either convex or nonconvex.

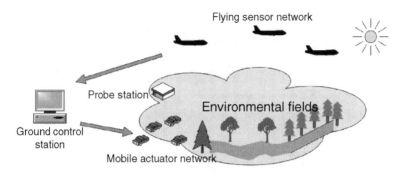

Figure 1.7 Cyber-physical systems.

A series of density functions $\varrho_1, \varrho_1, \varrho_3, \ldots$ are defined as $\varrho_i(q, t) \in [0, \infty)$, $\forall q \in \Omega$. For instance, ϱ_i could be wind direction, surface temperature, soil moisture level, and the like. The goal of monitoring a spatial–temporal process is to find the distribution of the required density functions:

$$\varrho_1(q, t), \varrho_2(q, t), \varrho_3(q, t), \ldots \quad \forall q \in \Omega, \forall t \in [t_1, t_2],$$

with preset spatial and temporal resolutions. The concept of using mobile sensor and actuator network to finish the remote sensing and distributed control missions is shown in Fig. 1.7. For example, a flying sensor network is sent out to collect the information of environmental fields. The ground probe station can be used for sensing validation. A group of ground robots serve as the actuator network to achieve the control missions. The whole system can also be called a cyber-physical system.

There are many challenges to realize this mission, especially for water, agriculture, or environmental applications, as stated in the following:

- *Low-Cost Solutions*: Most civilian applications are highly constrained by the cost. Sometimes, even satellite images are not affordable in a routine way for farmers.

- *Large-Scale Sensing*: Most water lands, agricultural fields are large-scale systems as big as tens or hundreds square miles.

- *High Temporal Requirements*: Some applications require the collections of images while the day light is as strong as possible.

- *High Spatial Requirements*: Many scenarios such as vegetation classification need much higher resolution image than what the normal satellite image can provide.

- *Easy Manipulation*: The civilian applications require the data collection procedure to be as easy as possible and as simple as possible.

- *Advanced Sensor/Actuator Allocation Algorithms*: Both the slow- and fast-evolving processes require properly designed strategies for sensing and control missions.

The above challenges could also be summarized as:

- What to measure and control? What entity needs to be measured and controlled?
- How to measure and control? What sensing or actuation systems are needed given a specific mission?
- When to measure and control? What time or how frequently to perform it?
- Where to measure and control? How to plan the trajectories of mobile sensors and actuators?

1.2   RESEARCH MOTIVATIONS

Several research motivations are listed in the following sections for the sensing and actuation missions of environmental or agricultural fields.

1.2.1   Small Unmanned Aircraft System Design for Remote Sensing

Many farms and environmental fields are tens or hundreds of square miles big. It is too expensive to build ground probes for monitoring purposes in all the interested areas. Remote sensing can provide a good solution here because of the wider sensing footprint from the air. There are several options for remote sensing including satellites, manned aircrafts, and UAVs. Satellite images can cover a large area, but they have low spatial resolutions and slow update rates. Manned aerial imagery is expensive to acquire. Small UAVs cost less money and can provide more accurate information from low altitudes with less interference from clouds. Small UAVs combined with ground and orbital sensors can even form a multiscale remote sensing system.

Technological advances in wireless networks and microelectromechanical systems make it possible to capture aerial images using small and inexpensive UAVs for civilian applications [1]. Small UAVs have a relatively short wingspan and light weight. They can be operated by one to two people [2]. Many of them can even be hand-carried and hand-launched. In fact, small UAVs are designed to fly at low altitude (normally less than 1000 m above the ground) to provide a close observation of the ground objects and phenomena.

The problem of remote sensing using small UAVs can be divided into three subproblems.

1.2.1.1   UAS Integration Problem

The primary problem for using small UAVs in remote sensing is the system integration because there are no commercial off-the-shelf (COTS) inexpensive multispectral UAV solutions for civilian applications to the author's best knowledge. To achieve the autonomous navigation and image capture functions, the minimal UAV system requires the navigation subsystem (autopilot) and the image subsystem. However, the autopilot alone costs around $5000 or more [2]. The integration of the image subsystem to the UAV also requires special considerations not only for the space and weight limits but also for the postflight image georeferencing processes. Different imagers (red–green–blue (RGB), near infrared (NIR), or thermal band) may have special requirements. It is very challenging to integrate all the COTS units and open-source projects into a robust and fully functional UAS.

1.2.1.2 Image Collection Problem

The image collection problem is how to plan the waypoints or trajectory of the UAV or group UAVs so that the images could be captured in a fastest way. The problem can be defined as follows. Given a random area of interest Ω; UAVs with functions of altitude and speed keeping and waypoint navigation: speed v, flight height h; camera with specification: focal length F, image sensor pixel size $PS_h \times PS_v$, image sensor pixel pitch $PP_h \times PP_v$, the "camera shooting interval" t_{shoot}; the desired image resolution res; the control objective is

$$\min t_{flight} = g(\Omega, h, v, \{q_1, \ldots, q_i\}, t_{shoot}, res), \tag{1.1}$$

subject to $v \in [v_1, v_2]$, $h \in [h_1, h_2]$, $t_{shoot} = k \times t_{shoot_{min}}$, where t_{flight} is the flight time of the UAV for effective coverage, $g(\Omega, h, v, t_{shoot})$ is the function to determine the flight path and flight time for effective coverage, $\{q_1, q_2, \ldots, q_i\}$ is a set of preset UAV waypoints, and k is a positive integer. This solution of this optimal problem is also the endurance time of the unmanned system, since both the image acquisition and the UAV trajectory following need to be considered.

1.2.1.3 Image Registration and Postprocessing Problem

After the images are captured, the next problem is to correlate the images with the temporal and spatial data, which is also called georeferencing or image registration. The image information could be further processed to get the final environmental data such as soil temperature, vegetation distributions, and so on.

All three subproblems need to be addressed to design a cheap and robust small UAS for the remote sensing mission.

1.2.2 State Estimation for Small UAVs

The estimation accuracy of the UAV states in the air can greatly affect the UAV flight control performance and the later image georeferencing results. However, many required states are either not directly measurable (e.g., the orientation) or sampled slowly (e.g., the position) with current MEMS sensors. With the emergence of cheaper and smaller MEMS inertial measurement units (IMUs), it is necessary to develop our own state estimation algorithm to further reduce the system costs and to provide more flexibilities in post-flight georeferencing. The following states are indispensable for both the autonomous control and image registration:

- *Position States*: Many civilian GP receivers can only provide position updates at 4 Hz or less, which can introduce a big error if the UAV is moving very fast.
- *Orientation States*: The orientation data are not directly measurable with the current MEMS technology. It can only be estimated using advanced filters from gyros, accelerometers, magnetometers, and the like.

The estimation of the above states requires sophisticated nonlinear filter design techniques and extensive fight test validations for UAV navigation uses. Many researchers have looked into this problem with different hardware platforms. The extended Kalman filter is introduced for MNAV IMU as part of an open-source project [3]. Other researchers also take the speed measurement into consideration for more accurate acceleration estimation [4]. With the current trend of modularization and standardization in the UAV design, UAV developers can save a large amount of time by buying cheap COTS IMUs and configure them into a complete navigation system. Thus, a systematic procedure for the state filter design, tuning, and validation needs to be developed to support these cheap COTS IMUs. Especially, the Kalman filter must be carefully designed to avoid divergence [5].

1.2.3 Advanced Flight Control for Small UAVs

The UAV has an obvious safety advantage over a manned aircraft at extremely low altitudes (e.g., ~100 m above the ground) because the autopilot can be used for the autonomous navigation replacing the human pilot. The autopilot or flight control system plays a key role not only for the flight stability and navigation but also for sensor interpretation considerations [2]. In a remote surveillance task, the navigation performance of UAVs while flying horizontally could highly affect the georeferencing accuracy of the acquired aerial images. Small or micro-UAV autonomous flight can be easily affected by many factors:

- *Wind*: Wind gusts present a significant control challenge for low-mass airplanes.
- *Flight Altitude*: UAVs may need to fly at a broad range of altitudes for different missions.
- *Payload Variations*: A good UAV flight controller should be robust to payload variations so that it will not stall with little perturbation.
- *Manufacturing Variations and Modeling Difficulties*: Many research UAVs are built from remote controlled (RC) air frames, making it hard to get an accurate dynamic model.
- *Resource Limitations*: Small or micro-UAVs are also constrained by the onboard resources such as limited accuracy for onboard inertial sensors, limited computational power, limited size and weight, and the like.

All the above factors make it very important to design a robust and flexible flight controller. A lot of researchers have looked into the problem of UAV modeling and control. Open-loop steady-state flight experiments are proposed for the aileron-(roll rate) and elevator-(pitch rate) loop system identification (ID) [6]. But the open-loop system ID has to have special requirements on UAV flight stability, which limits the roll and pitch reference signals to be as small as 0.02 rad. UAV model ID experiments can also be performed with human operators controlling the UAVs remotely. Different types of autoregressive with exogenous input (ARX) models are identified while the UAV is flying in loiter mode [7]. Human operators could generate open-loop responses but it may be impossible for some specially designed reference like pseudo-random binary signals (PRBS). Other researchers also tried closed-loop system ID method on separate channels of unmanned helicopters [8–10]. In summary, it is still challenging to accurately model and control small low-cost UAVs. More work on new modeling and controlling techniques are needed.

1.2.4   Cooperative Remote Sensing Using Multiple UAVs

Because of the large-scale characteristics of most environmental fields, many applications may require remote sensing of a large land area (more than 30 square miles) within a short time (less than 1 hour). Acquisition of imagery on this geographic scale is difficult for a single UAV. However, groups of UAVs (which we refer to as "covens") can solve this problem because they can provide images from more spectral bands in a shorter time than a single UAV.

The following missions will need multiple UAVs (covens) operating cooperatively for remote sensing:

- Measure $\eta_1, \eta_2, \eta_3, \ldots$ simultaneously;
- Measure $\eta_i(q, t)$ within a short time.

To fulfill the above requirements, UAVs equipped with imagers having different wavelength bands must fly in some formation to acquire the largest number of images simultaneously. The reason for this requirement is that electromagnetic radiation may change significantly, even over a period of minutes, which in turn may affect the final product of remote sensing.

Many researchers have already looked into the problem of using multiple unmanned vehicles or robots for environmental monitoring problem. A model-free robot path planning algorithm is introduced for the coverage control problem of a density field [11]. The adaptive and singular value-based sampling algorithms are proposed for the ocean field sampling problem [12,13]. However, most of the reported efforts focus on the density field instead of a more complex vector field. Rotary wing UAVs are also used for heat flux estimation with user customized pressure sensing unit [14]. To achieve the optimal measurement of the wind field, groups of UAVs could fly in formations for faster estimation [15]. It is still an open research problem on how to optimally employ groups of unmanned vehicles for specific remote sensing missions.

1.2.5   Diffusion Control Using Mobile Actuator and Sensor Networks

The monitoring and control of a diffusion process can be viewed as an optimal sensor/actuator placement problem in a distributed system [16]. Basically, a series of desired actuator positions are generated based on centroidal Voronoi tessellations (CVT) and later integrated with PID controllers for neutralizing control based on Voronoi partitions. CVT algorithm provides a non-model-based method for coverage control and diffusion control using groups of vehicles. The CVT algorithm is robust and scalable [17,18], and it can guarantee the groups asymptotically converging to the affected area even in multiple/mobile sources application [11].

1.3   MONOGRAPH CONTRIBUTIONS

The major contributions of this monograph include, but are not limited to, the following:

- Explained the system design and test of the AggieAir UAS, a low-cost multispectral remote sensing platform, which is one of the major contributions of this monograph work.
- Summarized several state estimation algorithms for attitude estimation using low-cost IMUs for small UAVs.
- Designed and implemented the very first fractional-order flight controller to guide the UAV in real flights.
- Jointly developed the path planning algorithm for the remote sensing mission using a single UAV with different ground resolution requirements.
- Implemented and validated the multivehicle consensus algorithm on MASnet platform with different communication topologies.
- Proposed the consensus CVT-based path planning algorithm for the diffusion control problem.

Most of the algorithms and theories are validated on the following simulation or experimental platforms:

- Aerosonde UAV software simulation platform for flight controller design, state estimation algorithm tests;
- AggieAir UAV experimental platform for remote sensing, flying sensor network;
- MASnet hardware simulation platform for formation control, consensus validations;
- DiffMAS2D software simulation platform for diffusion control simulation.

1.4 MONOGRAPH ORGANIZATION

The monograph is organized as the following. The research motivations and monograph contributions are introduced in Chapter 1. Chapter 2 is dedicated to the introduction of AggieAir UAS platform, a low-cost multispectral remote sensing platform, with detailed explanation on the system design requirements, subsystem structures, and flight test protocol developments. Chapter 3 focuses on the state estimation problems for small UAVs. The fractional order PI^α controller is designed and implemented on the roll-channel in Chapter 4 including model ID methods, controller designing procedure, and simulation/experimental validation results. Chapter 5 explains how to finish a typical remote sensing task using a single UAV with lots of application scenarios such as water area, farmland, road coverage, and the like. The remote sensing problem using multiple unmanned vehicles is presented in Chapter 6. The multivehicle consensus algorithm is first tested on the MASnet platform and then a new algorithm for wind profiling measurement using multiple UAVs is proposed and tested in the simulation. Chapter 7 is devoted to the distributed control of a diffusion process using mobile sensor and actuator networks. The consensus and CVT-based path planning algorithm is proposed and tested in simulations. Chapter 8 is for the conclusions and some future research suggestions.

REFERENCES

1. D. W. Casbeer, S. M. Li, R. W. Beard, T. W. McLain, and R. K. Mehra. Forest fire monitoring with multiple small UAVs. In *Proceedings of the American Control Conference*, pages 3530–3535, June 2005.

2. H. Chao, Y. Cao, and Y. Q. Chen. Autopilots for small unmanned aerial vehicles: a survey. *International Journal of Control, Automation, and Systems*, 8(1):36–44, 2010.

3. J. S. Jang and D. Liccardo. Small UAV automation using MEMS. *IEEE Aerospace and Electronic Systems Magazine*, 22(5):30–34, 2007.

4. D. B. Kingston and R. W. Beard. Real-time attitude and position estimation for small UAVs using low-cost sensors. In *Proceedings of the AIAA 3rd Unmanned Unlimited Systems Conference and Workshop*, number AIAA-2007-6514, Chicago, IL, September 2004.

5. L. Perea, J. How, L. Breger, and P. Elosegui. Nonlinearity in sensor fusion: divergence issues in EKF, modified truncated SOF, and UKF. In *Proceedings of the AIAA Guidance, Navigation and Control Conference and Exhibit*, number AIAA-2007-6488, August 2007.

6. J. Nino, F. Mitrachea, P. Cosynb, and R. D. Keyser. Model identification of a micro air vehicle. *Journal of Bionic Engineering*, 4(4):227–236, 2007.

7. H. Wu, D. Sun, K. Peng, and Z. Zhou. Modeling identification of a micro air vehicle in loitering flight based on attitude performance evaluation. *IEEE Transactions on Robotics and Automation*, 20(4):702–712, 2004.

8. Y. Lee, S. Kim, and J. Suk. System identification of an unmanned aerial vehicle from automated flight tests. In *Proceedings of the AIAA's 1st Technical Conference and Workshop on Unmanned Aerospace Vehicles*, May 2002.

9. G. Cai, B. M. Chen, K. Peng, M. Dong, and T. H. Lee. Modeling and control of the yaw channel of a UAV helicopter. *IEEE Transactions on Industrial Electronics*, 55(9):3426–3434, 2008.

10. S. Duranti and G. Conte. In-flight identification of the augmented flight dynamics of the RMAX unmanned helicopter. In *Proceedings of the Seventeenth IFAC Symposium on Automatic Control in Aerospace*, June 2007.

11. J. Cortés, S. Matínez, T. Karatas, and F. Bullo. Coverage control for mobile sensing networks. *IEEE Transactions on Robotics and Automation*, 20(20):243–255, 2004.

12. A. Caiti, G. Casalino, E. Lorenzi, A. Turetta, and R. Viviani. Distributed adaptive environmental sampling with AUVs: cooperation and team coordination through minimum-spanning-tree graph searching algorithms. In *Proceedings of the Second IFAC Workshop on Navigation, Guidance and Control of Underwater Vehicles*, April 2008.

13. D. O. Popa, A. C. Sanderson, V. Hombal, R. J. Komerska, S. S. Mupparapu, D. R. Blidberg, and S. G. Chappel. Optimal sampling using singular value decomposition of the parameter variance space. In *Proceedings of the IEEE/RSJ International Conference on Intelligent Robots and Systems*, pages 3131–3136, August 2005.

14. J. Bange, P. Zittel, T. Spieß, J. Uhlenbrock, and F. Beyrich. A new method for the determination of area-averaged turbulent surface fluxes from low-level flights using inverse models. *Boundary-Layer Meteorology*, 119(3):527–561, 2006.

15. Y. Q. Chen and Z. Wang. Formation control: a review and a new consideration. In *Proceedings of the IEEE/RSJ International Conference on Intelligent Robots and Systems*, pages 3181–3186, August 2005.

16. D. Ucinski. *Optimal Measurement Methods for Distributed Parameter System Identification*. CDC Press, Boca Raton, FL, 2004.

17. Y. Q. Chen, Z. Wang, and J. Liang. Actuation scheduling in mobile actuator networks for spatial-temporal feedback control of a diffusion process with dynamic obstacle avoidance. In *Proceedings of the IEEE International Conference on Mechatronics and Automation*, pages 1630–1635, June 2005.

18. H. Chao, Y. Q. Chen, and W. Ren. A study of grouping effect on mobile actuator sensor networks for distributed feedback control of diffusion process using central voronoi tessellations. *International Journal of Intelligent Control Systems*, 11(2):185–190, 2006.

Chapter 2

AggieAir: A Low-Cost Unmanned Aircraft System for Remote Sensing

2.1 INTRODUCTION

This chapter focuses on the design and testing of AggieAir, a small low-cost unmanned aircraft systems (UAS). The primary design purpose of AggieAir is for remote sensing of meteorological and/or other related conditions over agricultural fields or other environmentally important land areas. Small UAS, including unmanned aerial vehicle (UAV) with payload and ground devices, have many advantages in real-world remote sensing applications over traditional aircraft- or satellite-based platforms or ground-based probes. This is because small UAVs are easy to manipulate, cheap to maintain, and able to remove the need for human pilots to perform tedious or dangerous tasks. Multiple small UAVs can be flown in a group and accomplish challenging tasks such as real-time mapping of large-scale agriculture areas.

The purpose of remote sensing is to acquire information about the Earth's surface without coming into contact with it. One objective of remote sensing is to characterize the electromagnetic radiation emitted by objects [1]. Typical divisions of the electromagnetic spectrum include the visible light band (380–720 nm), near-infrared (NIR) band (0.72–1.30 μm), and mid-infrared (MIR) band (1.30–3.00 μm). Band-reconfigurable imagers can deliver images from different bands ranging from visible spectra to infrared or thermal band depending on specific applications. Different combinations of spectral bands can have different purposes. For example, the combination of the red band and infrared band can be used to detect vegetation and camouflage while the combination of red slope can be used to estimate the percent of vegetation cover [2]. Different bands of images acquired remotely through UAS could be used in scenarios such as water management and irrigation control [3]. In fact, it is difficult to sense and estimate the state of a water system because most water systems are large-scale and need monitoring of many factors including the quality, quantity, and

Remote Sensing and Actuation Using Unmanned Vehicles, First Edition. Haiyang Chao and YangQuan Chen.
© 2012 by The Institute of Electrical and Electronics Engineers, Inc.
Published 2012 by John Wiley & Sons, Inc.

location of water, soil, and vegetation. For the mission of accurate sensing of a water system, ground probe stations are expensive to build and can only provide data with very limited sensing range (at specific locations and second-level temporal resolution). Satellite images can cover a large area, but have a low-spatial resolution and a slow update rate (30–250 meter, or lower-spatial resolution and week-level temporal resolution). Small UAVs cost less money but can provide more accurate information (meter or centimeter spatial resolution and hour-level temporal resolution) from low altitudes with less interference from clouds. Small UAVs combined with ground and orbital sensors can even form a multiscale remote sensing system.

UAVs have been used in several agricultural remote sensing applications for collecting aerial images. High-resolution red–green–blue (RGB) aerial photos can be used to determine the best harvest time of wine grapes [4]. Multispectral images are also shown to be potentially useful for monitoring the ripeness of coffee [2]. Images from reconfigurable bands taken simultaneously can increase the final information content of the imagery and significantly improve the flexibility of the remote sensing process. However, most current UAV remote sensing applications use large and expensive UAVs with heavy cameras (one or several kilograms), which makes it impossible for frequent civilian uses. Water management is still a new application area for UAVs, but it has more precision requirements compared with other remote sensing applications. For example, real-time management of water systems requires lots of precise information on water, soil, and plant conditions with preset spatial and temporal resolutions.

Motivated by the above remote sensing problem, AggieAir™, a band-configurable small UAS-based remote sensing system, has been developed step-by-step at Center for Self-Organizing and Intelligent Systems (CSOIS) together with Utah Water Research Lab (UWRL), Utah State University. The objective of this chapter is to present an overview of the ongoing research efforts on AggieAir platform. Particularly, this chapter focuses more on the system-level design, integration, and testing with brief introductions on subsystem details.

The chapter first presents a brief overview of the UAS focusing on the core of the whole system: the autopilot. After introducing the common UAS structure, the hardware and software aspects of the autopilot control system are then explained. Different types of available sensor suites and autopilot control techniques are summarized. Several typical commercial off-the-shelf and open-source autopilot packages are compared. The chapter then introduces AggieAir, a small and low-cost UAS for remote sensing applications. AggieAir comprises a flying-wing airframe as the test bed, the OSAM-Paparazzi autopilot for autonomous navigation, the GhostFoto imaging system for image capture, the Paparazzi ground control station (GCS) for real-time monitoring, and the geospatial real-time aerial image display (gRAID) software for image postprocessing. AggieAir UAS is fully autonomous, easy to manipulate, and independent of a runway. AggieAir can carry embedded cameras with different wavelength bands, which are low-cost but have high spatial resolution. These imagers mounted on UAVs can form a camera array to perform multispectral imaging with reconfigurable bands, depending on the objectives of the mission. Developments of essential subsystems, such as the airframe, the UAV autopilot, the imaging payload

Table 2.1 Small UAS Categories

Group	Gross Take-Off Weight
i	≤4.4 lb or 2 kg
ii	≤4.4 lb or 2 kg
iii	≤19.8 lb or 9 kg
iv	≤55 lb or 25 kg
v	Lighter than air (LTA) only

subsystem, and the image processing subsystem, are introduced together with some typical UAVs developed and built at CSOIS.

2.2 SMALL UAS OVERVIEW

In this monograph, the acronym UAV (unmanned aerial vehicle) is used to represent a power-driven, reusable airplane operated without a human pilot on board. The UAS (unmanned aircraft system) is defined as a UAV and its associated elements, which may include ground control stations, data communication links, supporting equipments, payloads, flight termination systems, and launch/recovery equipments [5]. Small UAS (sUAS) could be categorized into five groups based on gross take-off weight by the sUAS aviation rule making committee [5], as shown in Table 2.1. Group *i* includes those constructed in a frangible manner that would minimize injury and damages if there is a collision, compared with the group *ii*.

The research and development of small UAS has been quite active in the past few years [6]. Many small fixed-wing or rotary-wing UAVs are flying in the air under the guidance from the autopilot systems for different applications like forest fire monitoring, coffee field survey, search and rescue, and the like. A typical UAS includes the following:

- *Autopilot*: An autopilot is a microelectromechanical system (MEMS) used to guide the UAV without assistance from human operators, consisting of both hardware and its supporting software. The autopilot is the base for all the other functions of the UAS platform.

- *Airframe*: The airframe is where all the other devices are mounted including the frame body, which can be made from wood, foam, metal, or composite materials. The airframe also includes the flight control surfaces, which could be a combination of either aileron/elevator/rudder, or elevator/rudder, or elevons.

- *Payload*: The payload of UAS could be cameras of different spectral bands, or other emission devices such as LIDAR mostly for intelligence, surveillance, and reconnaissance purposes.

- *Communication Subsystem*: Most UASs have more than one wireless link supported. For example, RC link for safety pilot, WiFi link for large data sharing, and data link for ground monitoring.

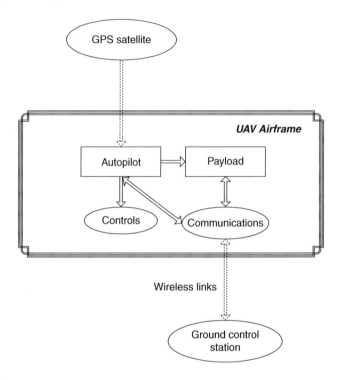

Figure 2.1   UAS structure.

- *Ground Control Station*: Ground control station is used for real-time flight status monitoring and flight plan changing.
- *Launch and Recovery Devices*: Some UAS may need special launching devices such as a hydraulic launcher or landing devices like a net [7].

The whole UAS structure is shown in Fig. 2.1. The minimal UAS onboard system requires the airframe for housing all the devices, the autopilot for sensing and navigation, the basic imaging payload for aerial images, and the communication subsystems for data link with the ground. The autopilot overview is focused first since it is the base for the UAS navigation and further image georeferencing.

Autopilot systems are now widely used in modern aircrafts and ships. The objective of UAV autopilot systems is to consistently guide UAVs to follow reference paths, or navigate through preset waypoints. A powerful UAV autopilot system can guide UAVs in all stages of autonomous flight including take-off, ascent, descent, trajectory following, and landing. An autopilot needs also to communicate with the ground station for control mode switch, to receive broadcasts from GPS satellites for position updates, and to send out control commands to the servo motors on UAV.

A UAV autopilot system is a closed-loop control system with two fundamental functions: state estimation and control signal generation based on the reference paths and the current states. The most common state observer is the inertial measurement

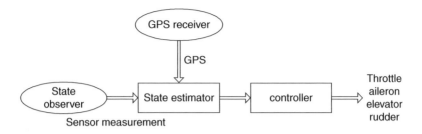

Figure 2.2   Functional structure of the UAV autopilot.

unit (IMU) including gyros, accelerometers, and magnetic sensors. There are also other attitude determination devices available such as infrared or vision-based ones. The sensor readings combined with the GPS information can be passed to a filter to generate the estimates of the current states for later control uses. Based on different control strategies, the UAV autopilots can be categorized as PID-based autopilots, fuzzy-based autopilots, neutral network (NN)-based autopilots, and so on. A typical commercial off-the-shelf UAV autopilot system comprises a GPS receiver, an IMU, and an onboard processor (state estimator and flight controller) as illustrated in Fig. 2.2.

2.2.1 Autopilot Hardware

The hardware of a minimal autopilot system includes sensor packages for state determination, onboard processors for estimation and control uses, and some peripheral circuits for servo and modem communications. Due to the physical limitations of small UAVs, the autopilot hardware must be small, light, and have a long endurance time. Moreover, the sensor packages should guarantee precise and robust performances, especially in mobile and temperature-varying outdoor environments.

2.2.1.1 MEMS Inertial Sensors

Inertial sensors are used to measure the 3D position and attitude information. The current MEMS technology makes it possible to use tiny and light sensors on small or micro UAVs. Available MEMS inertial sensors include the following:

- *GPS Receiver*: To measure the positions (p_n, p_e, h) and ground velocities (v_n, v_e, v_d).
- *Rate Gyro*: To measure the angular rates (p, q, r).
- *Accelerometer*: To measure the accelerations (a_x, a_y, a_z). Accelerometers actually measure the acceleration minus the gravity.
- *Magnetometer*: To measure the magnetic field for the heading correction (ψ).
- *Pressure Sensor*: To measure the air speed (the relative pressure) and the altitude (h).

- *Ultrasonic Sensor or SONAR*: To measure the relative height above the ground.
- *Infrared Sensor*: To measure the attitude angles (ϕ, θ).
- *RGB Camera or Other Image Sensors*: To replace one or several of the above sensors.

GPS receiver plays an indispensable role in the autonomous control of UAVs because it provides an absolute position measurement. A known bounded error between GPS measurement and the real position can be guaranteed as long as there is a valid 3D lock. For instance, u-blox 5 GPS receiver can achieve a 3-m 3D accuracy (pAcc for u-blox message) in the best case for civilian applications in the United States [8]. There are also differential GPS units that could achieve centimeter-level positioning accuracy. The disadvantage of GPS is its vulnerability to weather factors and its relatively low updating frequency (commonly 4 Hz or less for low-cost GPS receivers), which may not be sufficient for UAV applications.

2.2.1.2 Possible Sensor Configurations

Given all the above inertial sensors, several sensor combinations can be chosen for different types of UAVs to achieve the basic autonomous waypoints navigation task. Most current outdoor UAVs have GPS receivers onboard for position feedback. The major difference is the attitude measurement solution, which can be IMU, infrared sensor, image sensor, and so on.

Inertial Measurement Unit (IMU)   A typical IMU includes three-axis rate gyro and acceleration sensors, which can be filtered to generate an estimation of the attitude (ϕ, θ, ψ). A straightforward sensor solution for small UAVs is to use the IMU, which can provide a complete set of sensor readings. Microstrain GX2 is this kind of micro-IMU with an update rate up to 100 Hz for inertial sensing. It has three-axis magnetic, gyro, and acceleration sensors [9].

Infrared Sensor   Another solution for attitude sensing is using infrared thermopiles. The basic idea of infrared attitude sensor is to measure the heat difference between two sensors on one axis to determine the angle of the UAV because the Earth emits more IR than the sky. Paparazzi open-source autopilot group use infrared sensors as their primary attitude sensors [10,11]. However, we found that this solution is not accurate enough for postflight georeferencing purposes.

Vision Sensor   Vision sensor can also be used to estimate the attitude by itself or combined with other inertial measurements [12]. The pseudo roll and pitch can be decided from the onboard video or image streams [13]. Experiments on vision-only-based navigation and obstacle avoidance have been achieved on small rotary-wing UAVs [14]. In addition, vision-based navigation has potentials to replace the GPS for position measurements especially in task-oriented and feature-based applications. Vision-based navigation for small UAVs is still an undergoing topic and a lot of work is needed for mature commercial autopilots.

2.2.2 Autopilot Software

All the inertial measurements from sensors are sent to the onboard processor for further filtering and control processing. An autopilot could subscribe services from the available sensors based on different control objectives. The strength of the autopilot software directly affects the robustness of the whole autopilot system.

2.2.2.1 Autopilot Control Objectives

The UAV waypoints tracking task can be decomposed into several subtasks including

- pitch attitude holding
- altitude holding
- speed holding
- automatic take-off and landing
- roll-angle holding
- turn coordination
- heading holding

2.2.2.2 State Estimation

To achieve the above control objectives, different system states are needed with relatively high frequency (typically above 20 Hz for small fixed-wing UAVs). However, sensors like low-cost GPS can only provide a noisy measurement in 4 Hz. Kalman filter can be used here to make an optimal estimation (H_2) of the current states including the UAV location, velocity, and acceleration. The users need to define a noise estimation matrix, which represents how far the estimate can be trusted from the true states. Kalman filtering needs lots of matrix manipulations, which add more computational burdens to the onboard processor. Therefore, it is necessary to simplify the existing Kalman filtering techniques based on different applications. Besides, several other issues, such as gyro drifting and high-frequency sensor noise, also need to be canceled out through filtering techniques.

2.2.2.3 Controller Design for Autopilots

Most current commercial and research autopilots focus on GPS-based waypoints navigation. The path-following control of the UAV can be separated to different layers:

- inner loop on roll and pitch for attitude
- outer loop on heading and altitude for trajectory or waypoints tracking
- waypoint navigation

Kestrel has a 29-MHz Rabbit 3000 onboard processor with 512K RAM for on-board data logging. It has the built-in ability for autonomous take-off and landing, waypoint navigation, speed hold, and altitude hold. The flight control algorithm is based on the traditional PID control. The autopilot has elevator controller, throttle controller, and aileron controller separately. Elevator control is used for longitude and airspeed stability of the UAV. Throttle control is for controlling airspeed during level flight. Aileron control is used for lateral stability of the UAV [20]. Procerus provides in-flight PID gain tuning with real-time performance graph. The preflight sensor checking and failsafe protections are also integrated to the autopilot software package. Multiple UAV functions are supported by Kestrel autopilots via the ground ComBox.

2.2.3.2 Cloud Cap Piccolo

Piccolo family of UAV autopilots from Cloud Cap Company provides several packages for different applications. PiccoloPlus is a full featured autopilot for fixed-wing UAVs. Piccolo *II*, as shown in Fig. 2.4, is an autopilot with user payload interface added. Piccolo LT is a size optimized autopilot for small electric UAVs [22]. It includes inertial and air data sensors, GPS, processing, RF data link, and flight termination, all in a shielded enclosure [22]. The sensor package includes three gyros and accelerometers, one dynamic pressure sensor, and one barometric pressure sensor. Piccolo has special sensor configuration sections to correct errors like IMU to GPS antenna offset, avionics orientation with respect to the body frame.

Piccolo LT has a 40-MHz MPC555 onboard microcontroller. Piccolo provides a universal controller with different user configurations including legacy fixed-wing controller, neutral net helicopter controller, fixed-wing generation two controller, and PID helicopter controller. Fixed-wing generation two controller is the most commonly used flight controller for conventional fixed-wing UAVs. It includes support for altitude, bank, flaps, heading and vertical rate hold, and auto-take-off and landing.

Figure 2.4   Piccolo II autopilot [22].

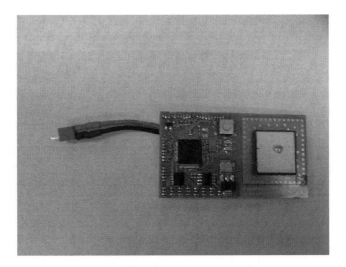

Figure 2.5 Paparazzi autopilot Tiny 13.

Piccolo autopilot supports one ground station controlling multiple autopilots and hardware-in-the-loop simulations.

2.2.3.3 Paparazzi Autopilot

Paparazzi autopilot is a popular project first developed by researchers from Ecole Nationale de l'Aviation Civile (ENAC) University, France. Infrared sensors combined with GPS are used as the default sensing unit. Although infrared sensors can only provide a rough estimation of the attitude, it is enough for a steady flight control once tuned well. Paparazzi Tiny 13 is the autopilot hardware with the GPS receiver integrated, shown in Fig. 2.5. Paparazzi also has Tiny without GPS (TWOG) autopilot with two open serial ports, which can be used to connect with IMU and modem. A Kalman filter runs on the autopilot to provide a faster position estimation based on GPS updates.

Paparazzi autopilot uses LPC 2148 ARM7 chip as the central processor. It can achieve waypoints tracking, auto-take-off and landing, and altitude hold. The flight controller can also be configured if a rate gyro is used for roll and pitch tracking on micro UAVs. However, Paparazzi autopilot does not have supports for speed holding currently since no air speed sensor reading is considered in the controller part. Paparazzi autopilot is also a truly autonomous autopilot without any dependence on the ground control station (GCS). It has many safety considerations such as exceptional handing in conditions such as RC signal lost, out of predefined range, GPS lost, and so on.

2.2.3.4 Specification Comparisons

The physical specifications of the autopilots are important since small UAVs demand as fewer space, payload, and power as possible. The size, weight, and power

Table 2.3 Comparison of Autopilot Functions

	Kestrel	Picocolo LT	Paparazzi TWOG
Waypints navigation	Y	Y	Y
Auto-take-off and landing	Y	Y	Y
Altitude hold	Y	Y	Y
Air speed hold	Y	Y	N
Multi-UAV support	Y	Y	Y
Attitude control loop	-	-	20/60 Hz
Servo control rate	-	-	20/60 Hz
Telemetry rate	-	25 Hz or faster	Configurable
Onboard log rate	≤100 Hz	-	N

consumption specifications are shown in Table 2.2. The functional specifications of these three typical autopilot are listed in detail in Table 2.3.

2.3 AGGIEAIR UAS PLATFORM

Most current autopilot systems can guide the UAVs to navigate through waypoints. However, it is not enough for real-world remote sensing applications because end users need aerial images with certain spatial and temporal resolution requirements. More importantly, most civilian remote sensing users require the UAV platform to be expendable. AggieAir UAS platform is developed considering all these remote sensing requirements. AggieAir is a small and low-cost UAV remote sensing platform, which includes the flying-wing airframe, the OSAM-Paparazzi autopilot, the GhostFoto image capture subsystem, the Paparazzi ground control station (GCS), and the gRAID software for aerial image processing. All the subsystems are introduced in this section.

2.3.1 Remote Sensing Requirements

Let $\Omega \subset R^2$ be a polytope including the interior, which can be either convex or nonconvex. A series of band density functions $\eta_{rgb}, \eta_{nir}, \eta_{mir}, \ldots$ are defined as $\eta_i(q, t) \in [0, \infty) \; \forall q \in \Omega$. η_{rgb} can also be treated as three bands η_r, η_g, η_b, which represent RED, GREEN, and BLUE band values of a pixel. The goal of remote sensing is to make a mapping from Ω to $\eta_1, \eta_2, \eta_3, \ldots$ with preset spatial and temporal resolutions for any $q \in \Omega$ and any $t \in [t_1, t_2]$ [3].

With the above remote sensing requirements, several specific characteristics need to be considered to get accurate georeferenced aerial images aside from an autonomous flying vehicle:

- *Expense*: Most civilian applications require expendable UAS platforms instead of expensive military grade unmanned vehicles. However, most commercial-off-the-shelf (COTS) autopilots cost more than $6000, let alone the camera and the air frame [23].

- *Orientation Data*: The orientation information when the image is taken is critical to the image georeferencing. Although the IR sensors can guide the UAV for autonomous flight, it is not enough for accurate georeferencing.

- *Image Synchronization*: Some COTS UAVs can send videos down to the base station for further processes. But there is a problem that the picture may not match up perfectly with the UAV data from the data log. The images may not synchronize perfectly with the orientation data from the autopilot.

- *Band Configurable Ability*: Many remote sensing applications require more than one band of aerial images such as vegetation mapping and some of them may require RGB, NIR, and even thermal images simultaneously.

2.3.2   AggieAir System Structure

AggieAir UAS includes the following subsystems:

- *The Flying-Wing Airframe*: Unicorn wings with optional 48, 60, and 72 in. wingspans are used as the frame bed to fit in all the electronic parts. The control actuators include elevons and a throttle motor.

- *The OSAM-Paparazzi Autopilot*: The open-source Paparazzi autopilot is modified by replacing the IR sensors with the IMU as the main sensing unit. Advanced navigation routines such as surveying an arbitrary polygon are also developed to support image acquisition of an area with a more general shape [24].

- *The GhostFoto Imaging Payload Subsystem [25]*: A high-resolution digital imaging subsystem with both the RGB and NIR bands is developed. More importantly, the image system could guarantee an accurate synchronization with the current autopilot software. COTS imaging payloads can also be fit on AggieAir UAVs including the analog video camera (VC) and the thermal camera.

- *The Communication Subsystem*: AggieAir has a 900-MHz data link for GCS monitoring, a 72-MHz RC link for safety pilot backup and an optional 2.4 GHz WiFi link for real-time image transmission.

- *The Paparazzi Ground Control Station (GCS)*: Paparazzi open-source ground station is used for the real-time UAS health monitoring and flight supervising.

- *The gRAID Software [24]*: A new NASA World Wind [26] plug-in named gRAID is developed for aerial image processing including correcting, georeferencing, and displaying the images on a 3D map of the world, similar to GoogleEarth.

There are two types of AggieAir UAVs developed and fully tested at CSOIS. AggieAir1 UAV is typically equipped with IR sensors for navigation and analog DV

Table 2.4   AggieAir2 UAS Specifications

	AggieAir2 Specifications
Take-off weight	Up to 8 lb
Take-off	Bungee
Wingspan	72 in.
Power	Lipo 11.1V-2000 mah $\times$ 8
Flight time	≤ 1 h
Cruise speed	15 m/s
Operational wind condition	< 10 m/s
Imaging payload	RGB/NIR/thermal camera
Payload weight	< 2 lb
Operational range	Up to 5 miles

for the image acquisition. AggieAir2 UAV is equipped with the commercial IMU for navigation and high-resolution GhostFoto imaging subsystem. The physical structure of AggieAir2 UAV is shown in Fig. 2.6, with the specifications in Table 2.4 and the airborne layout in Fig. 2.7 and Fig. 2.8.

AggieAir UAS has the following advantages over other UAS platforms for remote sensing missions:

- *Low Costs*: AggieAir airborne vehicles are built from scratches including the airframes and all the onboard electronics. The total hardware cost is less than $5000, which is the average cost of a COTS UAV autopilot.

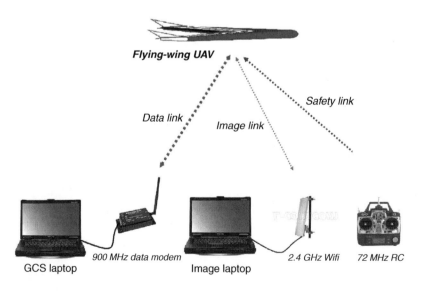

Figure 2.6   AggieAir UAS physical structure.

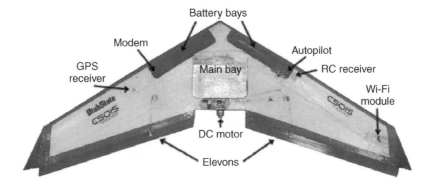

Figure 2.7 AggieAir2 UAS airborne layout.

- *Full Autonomy*: AggieAir uses the Paparazzi autopilot, which supports the full autonomy of the aerial vehicle even without the ground station.
- *Easy Manipulation*: Only two people are required to launch, manipulate, and land the vehicle.
- *Run-Way Free Capability*: The bungee launching system supports take-off and landing basically at any soft field with only one launching operator.

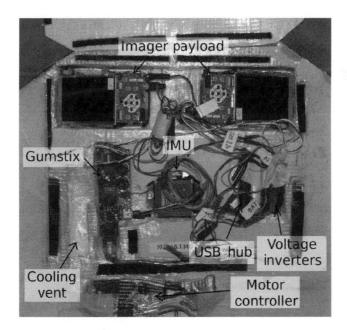

Figure 2.8 AggieAir2 main bay layout.

- *High Spatial and Temporal Resolutions*: The image system can achieve submeter-level ground resolution and hour-level time accuracy.

- *Multiple Spectral Bands for Imaging Payloads*: AggieAir supports RGB, NIR, and thermal bands for current image subsystems. More band configurable imagers are also under developments.

2.3.3 Flying-Wing Airframe

The airframe together with the actuators has a big impact on the UAV flight performance. It requires careful considerations to fit in all the sensors, actuators, and other subsystems. The flying-wing, or delta-wing, airframe is chosen as the research development platform due to its simplicity to build [20]. The flying-wing airframe is defined as the wings with no definite fuselage or tail. Theoretically, it is the most efficient aircraft design considering the aerodynamics and structural weight [27]. The Unicorn wing used in CSOIS includes 48, 60, and 72 in. wingspan [28], shown in Fig. 2.9, after gluing and taping.

The Unicorn flying-wing airframe has two control surfaces (left and right elevon) and one throttle motor to power the propeller. The Li-Poly batteries are chosen as the power sources due to their high-power density and light weights. Because of the current constraints from the batteries, both the throttle motor and the propeller need to be carefully selected to maximize the power efficiency while providing enough thrusts.

The major considerations while designing the airborne layout of the flying wings are to minimize the drag and choose the right center of gravity (CG). Most airborne parts such as the batteries and inner electronics are embedded inside the airframe and

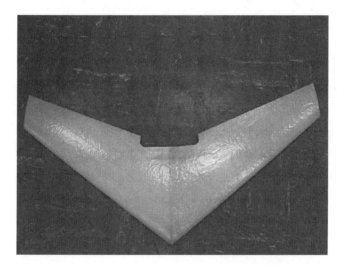

Figure 2.9 Unicorn wing.

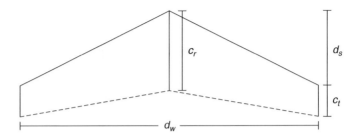

Figure 2.10   CG calculation for flying wings.

covered with plastic cases to prevent irregular flows, shown in Fig. 2.7. In addition, the aerodynamic shapes are designed for the cameras that have to reach out of the airframe. With the drag minimized, another very important design parameter is the position of the CG. The neutral point of all wings is at about 1/4 chord line [27]. The center of gravity of flying-wing airframes could be approximately calculated as follows [29]:

$$CG = \frac{d_s(c_r + 2c_t)}{3(c_r + c_t)} + b_p\% \left(c_r - \frac{2(c_r - c_t)(0.5c_r + c_t)}{3(c_r + c_t)} \right), \qquad (2.1)$$

where b_p is the percentage of the balance point on the mean aerodynamic chord, d_w is the wing span, c_r is the root chord, c_t is the tip chord, and d_s is the sweep distance. All the parameters above are defined in Fig. 2.10.

After the theoretical position of the CG is calculated, the layout can be designed to minimize the moment while turning. Most flying-wing airframes are designed to be a little nose heavy to have a positive pitching moment for the stability issues [30].

2.3.4 OSAM-Paparazzi Autopilot

It is clear that Kestrel and Piccolo autopilots are small, light, and powerful. But their prices are relatively high and most of their onboard software is not accessible to users, which is a main disadvantage when georeferencing aerial images after the flight [31]. Paparazzi UAV project provides a cheap, robust, and open-source autopilot solution including both hardware and software. But it uses infrared sensors for the attitude measurement, which is not accurate enough compared with most commercial UAV autopilots.

To achieve an accurate image georeferencing with a fair price, an inertial measurement unit (IMU) is added to the Paparazzi autopilot replacing the IR sensors. The detailed interface design is explained in later sections. Paparazzi TWOG board is used together with a 900-MHz Maxtream modem for real-time communication to the GCS. Microstrain GX2 IMU and u-blox LEA-5H GPS receiver serve as the attitude and position sensors, respectively. Due to the limits from the IO ports, the Gumstix microprocessor [32] is used as a bridge to connect IMU and GPS to the TWOG board. The cascaded PID flight controller then converts all the sensor information into PWM

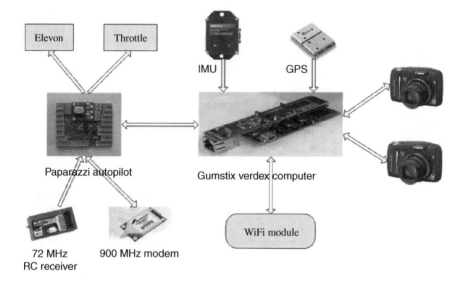

Figure 2.11   AggieAir airborne system structure.

signals for the elevon and throttle motor to guide the vehicle for preplanned navigation. There is also a 72-MHz RC receiver on board so that the human safety pilot
can serve as the backup for the autopilot in case of extreme conditions such as strong
wind gusts. The physical connections of the airborne system is shown in Fig. 2.11.

2.3.5   OSAM Image Payload Subsystem

To achieve the final goal of measuring the density functions of different bands simultaneously, two types of image subsystems have been developed including GhostFinger
and GhostFoto subsystems. GhostFinger subsystem is a stand-alone payload that
could either output streams of videos or trigger the picture capture by itself [24,33].
GhostFoto is a high-resolution imager that is connected with the IMU/GPS for more
accurate image georeferencing [25].

2.3.5.1   GhostFinger Subsystem

The weight, shape of the imager, and manner in which the imager is mounted
can significantly affect UAV flight dynamics. Due to the physical limitations of a
small UAV platform, the embedded imager must be small, light, and consume low
power. Four types of embedded imagers have been developed for small UAV applications: GhostFinger RGB digital camera (GF-RGB-DC), GhostFinger NIR digital
camera (GF-NIR-DC), GhostFinger RGB digital video camera (GF-RGB-DV), and
GhostFinger NIR digital video camera (GF-NIR-DV) [33].

Table 2.5   GhostFinger Digital Video Camera Specifications

Dimension:	22×16 mm
Weight:	≤ 100 g (including transmitter)
Power consumption:	About 600 mA at 5 V
Pixel size:	768×494
Focal length:	1.9 mm
Pixel pitch:	4.75 (h) $\times$ 5.55 (v) μm

Analog Video Camera   GF-DV is developed for real-time image transmission. It comprises a single-board analog 1/4 in CCD camera, one 2.4 GHz Black Widow wireless video transmitter, and peripheral circuits. The CCD camera only weighs about 7.09 g. The camera specifications are shown in Table 2.5 [33] and GF-DC is shown in Figs. 2.12 and 2.13 [33]. The communication range for the wireless transmitter and receiver can be up to 1 mile with the 3-dB antenna. The power consumption of the imager is estimated similarly with one 7.4 V Li-Poly battery and a voltage regulator. The GF-DV requires a wireless video receiver to get the video back in real-time.

RGB + NIR Digital Camera   The "GhostFinger digital camera" (GF-DC) is developed from a Pentax OPTIO E30 digital camera. Three photo capturing modes are included: timer triggered, RC triggered, and digital switch triggered. An ATMEG8L microcontroller is embedded with the camera to serve as a camera trigger controller. The timer trigger shot mode is for aerial photo capturing with a preset shooting interval t_{shoot}. The RC triggered mode is to use a spare channel from RC transmitter and receiver to control the picture capture manually via RC. The switch triggered shot mode is for triggering from the UAV autopilot to achieve a better geospatial

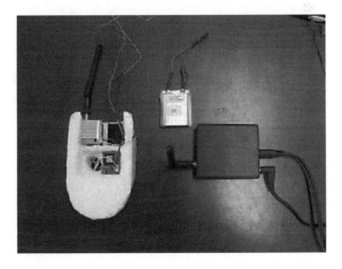

Figure 2.12   GhostFinger video camera [3].

Table 2.6 GhostFinger Digital Camera Specifications

Dimension:	6 in.
Weight:	174 g (battery not included)
Power consumption:	About 700 mA at 3.3V
Pixel size:	3072 × 2112
Focal length:	6–18 mm
Pixel pitch:	About 1.8 × 1.8 μm

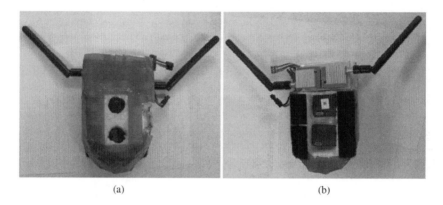

(a)   (b)

Figure 2.13 GhostFinger video camera 2 × 1 [3]. (a) Front view. (b) Back view.

synchronization. The camera specifications are shown in Table 2.6 [33] and GF-DC is shown in Figs. 2.14 and 2.15 [33]. The power consumption of the imager is estimated as an average value powered by one 3.7 V Li-Poly battery with 950 mA-h capacity. The images are saved on an SD card within the camera. The GF-DC has

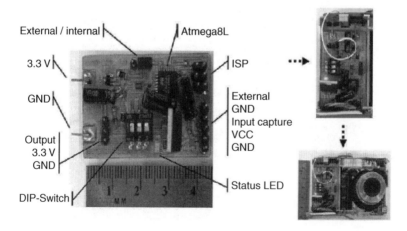

Figure 2.14 GhostFinger digital camera [33].

(a)   (b)

Figure 2.15 GhostFinger DC packages [33]. (a) Front view. (b) Back view.

been tested to work for more than 1 h in timer triggered mode with one 1GB SD card and a 3.7 V Li-Poly battery (950 mA-h).

Most CCD chips used on cameras are only sensitive to the electromagnetic light with a spectral wavelength ranging from 400 to 1100 nm, which includes both the visible band and NIR band. Digital cameras use an IR blocking filter to block wavelengths above 750 nm. A CCD-based camera can be changed into an NIR sensor by removing the IR blocking filter and adding one visual band blocking filter. A Lee 87 NIR filter is used in our GF-DC and GF-DV to block the light with the wavelength smaller than 730 nm. So, the GhostFinger NIR imager has a band of about 730–1100 nm. The NIR filter is added on both GF-DC and GF-DV to form a array of two imagers, which can measure the RGB and NIR simultaneously.

2.3.5.2 GhostFoto Imaging Subsystem

GhostFoto imaging subsystem is the second-generation remote controlled digital camera system developed at CSOIS [34]. The hardware includes the Canon CCD camera for image capture and the Gumstix microprocessor for payload control and georeferencing logging. Canon PowerShot SX100 IS CCD camera is used, shown in Fig. 2.16 [25]. This camera has the remote capturing capability, an 8-mega pixel CCD panel supporting up to 3264 by 2448 pixels size and a 10× optical zoom lens with optical image stabilizer. The compact size and relatively light weight (265 g) of this camera make it easy to fit the camera on small UAVs. Besides the commonly used RGB bands, the camera can also support near-infrared spectra by replacing the visible light filter with an NIR filter. Our 72 in. airframe can carry two or three of these imagers with different bands after removing unnecessary parts.

GFoto cameras are remotely controlled by the Gumstix through USB 1.1 interface with GhostEye image capture software. GhostEye is based on `libgphoto2` [35], which is an open-source portable digital camera library of C functions for UNIX-like operating systems. With `libgphoto2` library functions, GhostEye is able to remotely control and configure multiple cameras simultaneously through a Picture

Figure 2.16　Camera body (left) and its CCD sensor (right) [25].

Transfer Protocol (PTP) driver. PTP is a widely supported protocol developed by the International Imaging Industry Association for transfer of images from digital cameras to computers [36]. GhostEye also provides the communication link between the payload and the UAV system. Messages can be transmitted from GhostEye to the ground station, or can be shared among UAVs with the same protocol. Meanwhile, messages from the UAV system can trigger the imagers. For example, after the desired altitude is reached, the UAV is able to command the imager to activate or deactivate capturing. The georeferencing data is logged by GhostEye in XML format for easy imports to the gRAID [24].

2.3.6　gRAID Image Georeference Subsystem

The Geospatial Real-Time Aerial Image Display (gRAID) is a plug-in for NASA World Wind, serving as a 3D interactive open-source world viewer [24]. gRAID takes the raw aerial images, makes corrections for the camera radial distortion, and then overlays the images on the 3D Earth based upon the position and orientation data collected when they are captured. This process can be done either in real-time while the plane is flying or after the flight. Human-in-the-loop feature-based image stitching or mosaicing can be done with conventional GIS software after gRAID exports the image to a world file. gRAID can also create a gray scale image from a single RGB channel. The images can be converted into world files and loaded into conventional GIS software for further advanced image processing. The detailed georeferencing procedure is described as below.

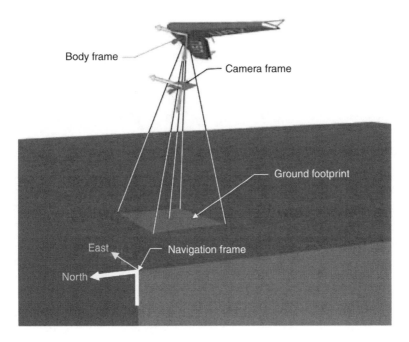

Figure 2.17 Aircraft coordinate systems.

To georeference the aerial images, several coordinate systems must first be defined, as shown in Fig. 2.17 [24].

- *The Body Frame*: The origin is defined at the center of gravity (CG), with the x-axis pointing through the nose, the y-axis pointing to the right wing, and the z-axis pointing down.

- *The Camera Frame*: The origin is located at the focal point of the camera. The axes of the camera frame are rotated by ϕ_c, θ_c, and ψ_c with respect to the body frame.

- *The Inertial Frame*: The origin is usually defined on the ground with the x, y, z axes pointing toward the north, east, and down, respectively. The orientation of the UAV with respect to the inertial frame is given by ϕ, θ, and ψ.

- *The Earth-Centered Earth-Fixed (ECEF) Frame*: The z-axis passes through the north pole, the x-axis passes through the equator at the prime meridian, and the y-axis passes through the equator at $90°$ longitude.

Any point in an image can be rotated from the camera frame to the ECEF coordinate system in order to find where it is located on the Earth. However, it is only necessary to find the location of the four corners of the image in order to georeference it. Assuming the origin is at the focal point and the image is on the image plane (2.2) can be used to find the four corners of the image. As defined in Fig. 2.18 [24], FOV_x

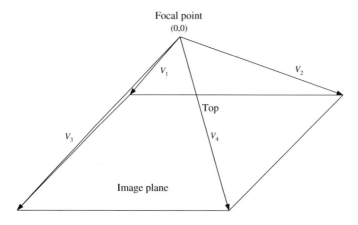

Figure 2.18   Definition of initial image corners.

is the FOV around the x axis, FOV_y is the field of view (FOV) around the y-axis, and f is the focal length,

$$v_c^1 = \left[\, f \tan(\text{FOV}_y/2) \; -f \tan(\text{FOV}_x/2) \; f \,\right],$$
$$v_c^2 = \left[\, f \tan(\text{FOV}_y/2) \; f \tan(\text{FOV}_x/2) \; f \,\right],$$
$$v_c^3 = \left[\, -f \tan(\text{FOV}_y/2) \; -f \tan(\text{FOV}_x/2) \; f \,\right],$$
$$v_c^4 = \left[\, -f \tan(\text{FOV}_y/2) \; f \tan(\text{FOV}_x/2) \; f \,\right]. \tag{2.2}$$

To rotate the corners into the navigation frame, they first need to be rotated to the body frame. The Euler angles with respect to the body frame are given by ϕ_c, θ_c, and ψ_c, and can be used to create a clock-wise rotation matrix R_c^b, which rotates a vector in the body frame to the camera frame,

$$R_c^b = R_{xyz}(\phi_c, \theta_c, \psi_c). \tag{2.3}$$

To rotate from the camera frame to the body frame, the transpose of R_c^b is used.

$$R_b^c = \left(R_c^b\right)^T = R_{zyx}(-\theta_c, -\psi_c, -\phi_c) \tag{2.4}$$

The same rotation matrix is used, with ϕ, θ, and ψ to rotate from the body frame into the navigation frame,

$$R_n^b = \left(R_b^n\right)^T = R_{zyx}(-\theta, -\psi, -\phi). \tag{2.5}$$

Now each corner is rotated from the camera frame into the navigation frame using (2.6),

$$v_n^i = R_n^b R_b^c v_c^i. \tag{2.6}$$

Since the corners are in the NED coordinate system, they can be scaled to the ground to find their appropriate magnitude (assuming flat Earth) where h is the height of the UAV above the ground and $v_n^i(z)$ is the z component of v_n^i.

$$v_n^i = v_n^i \frac{h}{v_n^i(z)} \qquad (2.7)$$

The next step is to rotate the image corners into the ECEF coordinate system. This is done with another rotation matrix and the latitude (λ) and longitude (α) of the UAV, that is,

$$R_w^n = R_{zyy}\left(-\alpha, \frac{\pi}{2}, \lambda\right), \qquad (2.8)$$

$$v_w^i = R_w^n v_n^i. \qquad (2.9)$$

After the corners are rotated into the ECEF coordinate system, they are located in the center of the Earth and need to be translated up to the position of the UAV in Cartesian coordinates (p),

$$v_w^i = v_w^i + p. \qquad (2.10)$$

So, v_w^i now represents the position of each of the image corners, in Cartesian coordinates, projected on the Earth.

2.4 OSAM-PAPARAZZI INTERFACE DESIGN FOR IMU INTEGRATION

One of our major breakthroughs of OSAM-Paparazzi autopilot is the integration of the IMU to the Paparazzi autopilot. The advantages to replace the default IR sensors include more accurate flight performance, easier controller tuning, and more accurate image georeferencing results. Due to the price and time requirements, the stand-alone IMU (Microstrain GX2) with orientation outputs is chosen for the AggieAir2 UAV instead of the IMU with GPS integrated (e.g., Xsens Mti-g). The Microstrain GX2 IMU has a RS232 interface [9] and the GPS receiver has a UART interface. With the above hardware, the following constrains for the OSAM-Paparazzi interface design must be considered:

- The interface needs to collect data from the GPS and IMU both through serial ports.
- The only two UARTs of Paparazzi autopilot are already used by GPS and data modem.
- The digital cameras need one more processor for remote control and image logging.
- The onboard devices need to be light, small, and stable to put on the UAV.
- The hardware and software modifications should be minimized due to the time constrains.

Table 2.7   Serial Mapping of Gumstix Console-VX Board

Dev ID	UART ID	DIN8 Position	Default Function	Max Rate (kbps)
/dev/ttyS0	FFUART	Middle	Default Linux Console	230
/dev/ttyS1	BTUART	Upper	Blue Tooth Comm.	921
/dev/ttyS2	STUART	Corner	General Purpose	230

- The onboard devices need to be powerful for later onboard image processing and inter-UAV communication purposes.

2.4.1   Hardware Interface Connections

Based on the above constrains, mostly the IO limits, the Gumstix is chosen to serve as both the imager controller and the bridge to connect the sensors to the Paparazzi autopilot. The Gumstix Verdex Pro XM4 board has a 400-MHz PXA270 processor and 64M RAM with the embedded Linux system [32]. The Verdex Pro mother board is connected with a Console-VX board for extra IOs including three RS232 Mini-DIN8 ports and one USB port, shown in Fig. 2.19. There are also three UART ports on the Console-VX board, which can take in TTL logic level (3.3V) signals, shown in Table 2.7.

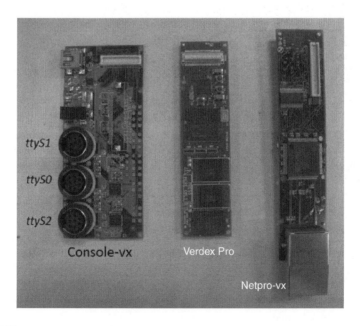

Figure 2.19   Gumstix Verdex and Console-VX boards.

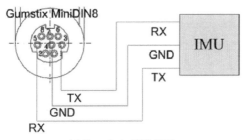

(a) Gumstix to GX2 IMU.

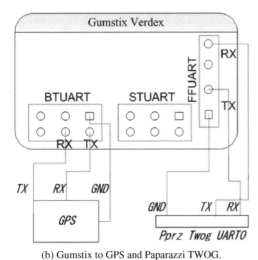

(b) Gumstix to GPS and Paparazzi TWOG.

Figure 2.20   OSAM-Paparazzi hardware connection.

The Microstrain GX2 IMU is connected with the Gumstix through the Mini-DIN8 port, shown in Fig. 2.20a. Both the GPS module and the Paparazzi autopilot are connected through the UARTs, shown in Fig. 2.20b.

2.4.2   Software Interface Design

Given the above hardware, two interface software programs are developed. They are (1) GhostGX2 interface running on Gumstix, and (2) OSAM-IMU interface running on Paparazzi autopilot. Both interfaces are written in C for compatibility and efficiency considerations. The GhostGX2 interface has the following functions (in priorities):

- Collect data in bytes from the IMU and GPS in real-time.
- Parse the data when the whole packet is ready.
- Transmit the orientation, position, and velocity data to the Paparazzi autopilot in real-time for control purposes.
- Control the onboard cameras remotely when needed.

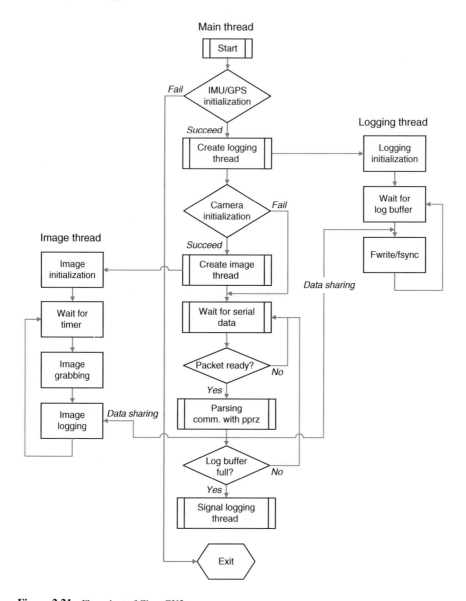

Figure 2.21   Flow chart of GhostGX2.

- Share the orientation and position data with the image part for later georeferencing purposes.
- Transmit the GPS accuracy data and satellite data to the Paparazzi autopilot for monitoring purposes.
- Log all the raw sensor data on the Gumstix for emergency analysis and further data filtering.

Although the Gumstix Verdex has powerful computational abilities (400 MHz), fast IO manipulations can still slow down the program and may affect the real-time communication with the autopilot. Example IO manipulations include camera control through USB ports and real-time data logging through serial (50 or 100 Hz). Thus, GhostGX2 is designed to run in multiple threads to avoid the interference from the slow IOs. The image thread and the logging thread are singled out with only global variable sharing through memory copy in a slow speed (every second or several seconds). The program flow chart of GhostGX2 is shown in Fig. 2.21. The Algorithms 2.4.1 and 2.4.2 are designed to achieve the predefined functions. The communication protocols of several representing IMUs and GPS are provided in Appendix A. The implementation details of GhostGX2 are introduced in Appendix B. The imaging thread implementation is also similarly implemented [34].

The Paparazzi airborne code also needs to be modified slightly to be able to understand the data from the Gumstix. The designed OSAM-IMU interface on Paparazzi autopilot can achieve the following functions:

- Collect the orientation, position, velocity, and GPS data from the Gumstix.
- Collect the GPS accuracy and satellite information from the Gumstix.
- Parse the data when the packet is ready.
- Update the attitude and GPS variables in the memory.

The OSAM-IMU files written in C are added on the airborne codes with the program flow chart shown in Fig. 2.22. The whole data receiving and parsing parts are implemented in an interrupt or event task. The communication protocol

---

Algorithm 2.4.1. GhostGX2 Interface

```
while(1){
    ...
    select(imu, gps);                              /* Wait data from imu and gps */
    parse_data();
    if (packet_ready)
        osam_link_pprz();                          /* Transmit to Paparazzi. */
    ...
    pthread_mutex_lock(andmutex);                  /* Main thread locks log buff */
        write_global_log_buf();
        if (log_buf_ready)
            pthread_cond_signal(andlog_buf_ready); /* Signal the log thread */
    pthread_mutex_unlock(andmutex);
    ...
    pthread_mutex_lock(andmutex2);                 /* Main thread locks image buff */
        write_global_image_buf();
    pthread_mutex_unlock(andmutex2);
    ...
}
```

Algorithm 2.4.2. Real-Time Logging

```
while(no_error){
   ...
   pthread_mutex_lock(andmutex);                    /* Log thread locks log buff */
      pthread_cond_wait(andlog_buf_ready,andmutex);
      global_buf_sharing();
   pthread_mutex_unlock(andmutex);
   ...
   fwrite(andglobal_buf,log_file);
   fflush(log_file);
   fsync(log_file);                                 /* Time-consuming IO manipulations */
   ...
}
```

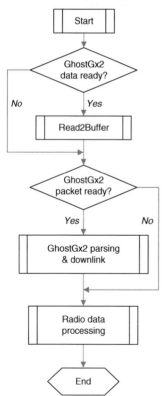

Figure 2.22   Flow chart of OSAM interface on Paparazzi autopilot.

Table 2.8 Communication Protocol Between Gumstix and Paparazzi

Bytes	1	2	3	4	5 ~ 4+LEN	5+LEN	6+LEN
	93	E0	ID	LEN	PAYLOAD	CHECKSUM1	CHECKSUM2

Table 2.9 Packet Definition for Data from GhostGX2

Packet ID	Length (byte)	Rate (Hz)	Description
0	38	4	GPS position, velocity, and accuracy
1	12	50	Orientation
2		1	GPS status and satellite inforormation
3		0.5	Image capturing information

between Gumstix and Paparazzi is defined in Table 2.8. The packet types received on the Paparazzi autopilot are defined in Table 2.9.

2.5 AGGIEAIR UAS TEST PROTOCOL AND TUNING

With the new GhostGX2 interface added to Paparazzi autopilot, it is very important to test the whole AggieAir system for both the functionality and robustness. Besides, many controller parameters and exceptional values need to be determined based on our specific UAV platform. Both the test protocol and the controller tuning procedure are developed in this book.

2.5.1 AggieAir UAS Test Protocol

To maximize the outcome of the flight test results and minimize the potential property damages, the AggieAir UAS Test Protocol is developed for the whole testing procedure with any airborne code modifications. A series of flight test preparation and manipulation documents are developed, shown in detail in the Appendix C. The whole code test procedure is as follows, shown in Fig. 2.23. First, each problem is analyzed and attacked separately. Then, the code is developed and verified on the ground. All the new functions are tested in the laboratory first before the flight test. Finally, the real flight tests serve as the feedback for the testing procedure.

One example of ground test is shown in Table 2.10. This ground test was for monitoring the memory and CPU usage of GhostGX2. Top is used to get the Gumstix resource information. GhostGX2 is left running for more than an hour with the camera capturing every 5 s. It can be seen that the memory used only increased slowly due to the slow speed of the data logging and the system is stable enough during the designed maximal time period.

Table 2.10   Gumstix Resource Monitoring for GhostGX2

No.	Time (h:mm)	Used Mem. (k)	Free Mem. (k)	CPU Usage (%)
1	4:29	21472	41452	N/A
2	4:34	22492	40432	N/A
3	4:44	24748	38176	N/A
4	4:50	25984	37940	N/A
5	4:54	26800	36124	N/A
6	4:59	27940	35004	N/A
7	5:04	28972	33952	7.6
8	5:12	30880	32044	7.6
9	5:17	31936	30988	7.6
10	5:27	34552	28372	7.6
11	5:34	35584	27340	7.6

2.5.2   AggieAir Controller Tuning Procedure

The flight control system of Paparazzi autopilot comprises several cascaded PID controllers, which need some tunings for every specific airframe. A well-tuned controller can maximize the flight time and minimize the probability of stalling. A special controller tuning procedure is developed here for our flying-wing airframes.

The key controller parameters of Paparazzi autopilot include [37]

- nominal_speed, the default ground speed
- cruise_throttle, the default throttle percentage in steady flights

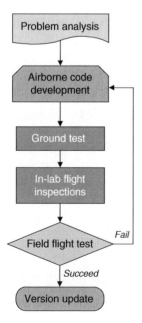

Figure 2.23   AggieAir UAS test protocol.

- default_circle_radius, the default turning radius
- roll_neutral, for the compensation on the roll channel caused by building asymmetries
- pitch_neutral, for the default angle of attack during steady flights
- roll_pgain, k_p of the roll loop P-controller
- pitch_pgain, k_p of the pitch loop P-controller
- course_pgain, k_p of the course loop PI-controller
- altitude_pgain, k_p of the altitude loop PI-controller.

Based on extensive flight test experiences, the following algorithm is summarized for the controller tuning of AggieAir UAV platforms.

Algorithm 2.1

The OSAM-UAV controller tuning procedure can be described as below:

(1) Set up default values for all the parameters based on RC flight performance;

(2) Trim tuning in manual mode to achieve the steady flight at the preset cruise_throttle;

(3) Pitch-loop rough tuning (pitch_neutral, pitch_pgain) to guarantee no altitude loss while flying straight in Auto1 mode;

(4) Roll-loop rough tuning (roll_neutral, roll_pgain) for the symmetry of left and right circle in Auto1 mode;

(5) Pitch-loop and roll-loop fine tuning while following the flight pattern of straight line following, left circling, and right circling (Auto2 mode);

(6) Course-loop tuning in Auto2 mode.

2.6 TYPICAL PLATFORMS AND FLIGHT TEST RESULTS

Both AggieAir1 and AggieAir2 UAV platforms have been built and fully tested at CSOIS for different purposes. The autonomous flight data, shown in the flight test results part, are to demonstrate the effectiveness and robustness of the AggieAir UAS.

2.6.1 Typical Platforms

Several AggieAir UAVs developed or tested by the author are shown in Table 2.11 with detailed specifications. One typical AggieAir1 UAV, "Pheonix" is shown in Fig. 2.24 mounted with the IR sensors and GF-DV. AggieAir2 UAV, "Snow" is

Table 2.11 UAVs Developed (or Partly Developed) by H. Chao

Name	Wing Span (in.)	Weight (lb)	Sensors	Imagers	RC Ch.	IP
Pheonix	48	4.4	IR	3 VC	59	N/A
ChangE	60	6.4	GX2 IMU	N/A	42	252
Falcon	72	About 7	Mti-g IMU	2 DC	59	107
Griffin	72	About 7	GX2 IMU	2 DC	50	120
Tiger	72	7.8	GX2 IMU	2 DC	59	140
PinkPanther	72	About 8	GX2 IMU	N/A	59	
Snow	72	4.4	GX2 IMU	Thermal	50	133

specially designed for the thermal camera application with a cover on the top for protection and aerodynamic considerations, shown in Fig. 2.25. Other than airframes made from foam, there is also one airframe, "PinkPanther," covered by fiber glass. Both the electronics and the autonomous tuning are finished by the author.

2.6.2 Flight Test Results

One example flight result is shown in the following from the 2009 AUVSI student UAS competition. The desired and real-flight path is shown in Fig. 2.26 and the

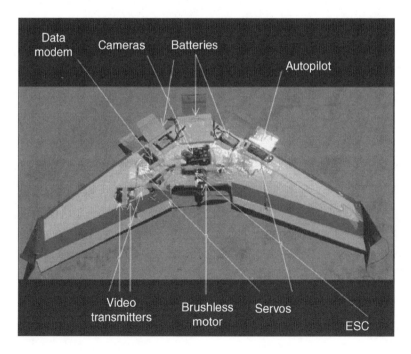

Figure 2.24 Layout of Pheonix 48 in. UAV.

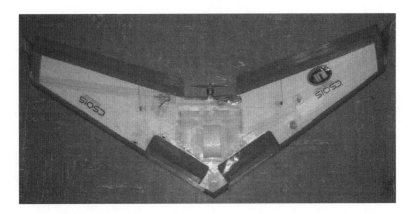

Figure 2.25 Layout of Snow 72 in. UAV.

altitude tracking results are shown in Fig. 2.27. It can be seen that the altitude error is roughly around 10 m, which is close to the system limit considering the UAV relies on GPS for the altitude control.

The failure rate during the preparation for the 2009 AUVSI student UAS competition is summarized in Table 2.12. The successful rate for autonomous navigation during 14 flight tests shows the effectiveness and robustness of the AggieAir UAS platform.

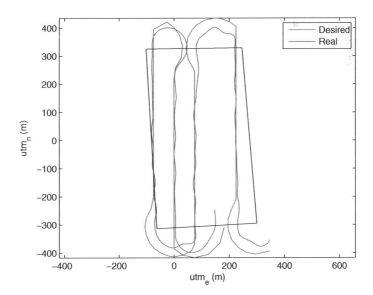

Figure 2.26 Sample flight trajectories of AggieAir2 UAV (Tiger).

Table 2.12   Robustness of AggieAir UAS

Flight Test Summary	
Number of flight tests	14
UAV flight time	14.3 h
Autonomous flight time	10.9 h
Success rate (auto-take-off)	99.4% (51/54)
Success rate (auto-landing)	98.1% (51/52)
Success rate (auto-navigation)	94.2% (49/52)

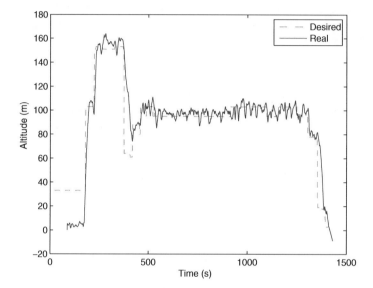

Figure 2.27   AggieAir altitude tracking results.

2.7   CHAPTER SUMMARY

In this chapter, the AggieAir1 and AggieAir2 UAV platforms are explained in detail from both the hardware and software levels. The current autopilot markets are investigated thoroughly based on the remote sensing requirements. The Paparazzi open-source autopilot is adapted to support the COTS IMU through our OSAM-Paparazzi interface. A series of testing protocols are proposed for flight safety considerations. Finally, some flight data are shown to prove the effectiveness of the whole system.

REFERENCES

1. B. C. James. *Introduction to Remote Sensing*. 4th ed. Guilford Press, New York, 2006.

2. L. F. Johnson, S. R. Herwitz, B. M. Lobitz, and S. E. Dunagan. Feasibility of monitoring coffee field ripeness with airborne multispectral imagery. *Applied Engineering in Agriculture*, 20:845–849, 2004.

3. H. Chao, M. Baumann, A. M. Jensen, Y. Q. Chen, Y. Cao, W. Ren, and M. McKee. Band-reconfigurable multi-UAV-based cooperative remote sensing for real-time water management and distributed irrigation control. In *Proceedings of the International Federal of Automatic Control (IFAC) World Congress*, pages 11744–11749, July 2008.

4. L. F. Johnson, S. R. Herwitz, S. E. Dunagan, B. M. Lobitz, D. V. Sullivan, and R. E. Slye. Collection of ultra high spatial and spectral resolution image data over California vineyards with a small UAV. In *Proceedings of the International Symposium on Remote Sensing of Environment*, November 2003.

5. B. Tarbert, T. Wierzbanowski, E. Chernoff, and P. Egan. Comprehensive set of recommendations for sUAS regulatory development. Technical Report, Small UAS Aviation Rulemaking Committee, 2009.

6. H. Chao, Y. Cao, and Y. Q. Chen. Autopilots for small unmanned aerial vehicles: a survey. *International Journal of Control, Automation, and Systems*, 8(1):36–44, 2010.

7. Insitu company. Insitu UAV introduction, 2010. http://www.insitu.com/.

8. u-blox company. u-blox GPS protocol, 2010. http://www.u-blox.com.

9. Microstrain Inc. Gx2 IMU specifications, 2008. http://www.mirostrain.com.

10. Open Source Paparazzi UAV Project, 2008. http://www.recherche.enac.fr/paparazzi/.

11. G. K. Egan. The use of infrared sensors for absolute attitude determination of unmanned aerial vehicles. Technical Report MECSE-22-2006, Monash University, 2006.

12. P. J. Roberts, R. A. Walker, and P. J. O'Shea. Fixed wing UAV navigation and control through integrated GNSS and vision. In *Proceedings of the AIAA Guidance, Navigation, and Control Conference and Exhibit*, number AIAA 2005-5867, August 2005.

13. D. Damien, B. Wageeh, and W. Rodney. Fixed-wing attitude estimation using computer vision based horizon detection. In *Proceedings of the Australian International Aerospace Congress*, pages 1–19, March 2007.

14. A. J. Calise, E. N. Johnson, M. D. Johnson, and J. E. Corban. Applications of adaptive neural-network control to unmanned aerial vehicles. In *Proceedings of the AIAA/ICAS International Air and Space Symposium and Exposition: The Next 100 Years*, June 2003.

15. H. Chao, Y. Luo, L. Di, and Y. Q. Chen. Fractional order flight control of a small fixed-wing UAV: controller design and simulation study. In *Proceedings of the ASME International Design Engineering Technical Conference Computers and Information in Engineering*, number MESA-87574, September 2009.

16. M. Kumon, Y. Udo, H. Michihira, M. Nagata, I. Mizumoto, and Z. Iwai. Autopilot system for Kiteplane. *IEEE/ASME Transactions on Mechatronics*, 11(5):615–624, 2006.

17. E. N. Johnson and S. Kannan. Adaptive flight control for an autonomous unmanned helicopter. In *Proceedings of the AIAA Guidance, Navigation and Control Conference*, number AIAA-2002-4439, August 2002.

18. M. Liu, F. Santoso, and G. K. Egan. Linear quadratic optimal control synthesis for a UAV. In *Proceedings of the 12th Australian International Aerospace Congress*, March 2007.

19. M. Sadraey and R. Colgren. 2 DOF robust nonlinear autopilot design for a small UAV using a combination of dynamic inversion and h-infinity loop shaping. In *Proceedings of the AIAA Guidance, Navigation and Control Conference*, number AIAA-2005-6402, August 2005.

20. R. Beard, D. Kingston, M. Quigley, D. Snyder, R. Christiansen, W. Johnson, T. Mclain, and M. Goodrich. Autonomous vehicle technologies for small fixed wing UAVs. *Journal of Aerospace Computing, Information, and Communication*, 5(1):92–108, 2005.

21. Procerus Technologies. Kestrel autopilot specifications, 2008. http://www.procerusuav. com.

22. CloudCap Inc. Piccolo autopilot specifications, 2008. http://www.cloudcaptech.com.

23. H. Chao, Y. Cao, and Y. Q. Chen. Autopilots for small fixed wing unmanned air vehicles: a survey. In *Proceedings of the IEEE International Conference on Mechatronics and Automation*, pages 3144–3149, August 2007.

24. A. M. Jensen. gRAID: A geospatial real-time aerial image display for a low-cost autonomous multispectral remote sensing platform. Master's thesis, Utah State Univeristy, 2009.

25. Y. Han. An autonomous unmanned aerial vehicle-based imagery system development and remote sensing images classification for agricultural applications. Master's thesis, Utah State Univeristy, 2009.

26. NASA. World wind software, 2007. http://worldwind.arc.nasa.gov/index.html.

27. W. F. Phillips. *Mechanics of Flight*. John Wiley & Sons, Hoboken, NJ, 2004.

28. The Unicorn Assembly and Flight Manual, 2004. http://http://unicornwings.stores. yahoo.net/.

29. Palos RC Flying Club. CG calculator, 2009. http://www.palosrc.com/instructors/cg.htm.

30. B. L. Stevens and F. L. Lewis. *Aircraft Control and Simulation*. 2nd ed. John Wiley & Sons, Hoboken, NJ, 2003.

31. A. M. Jensen, M. Baumann, and Y. Q. Chen. Low-cost multispectral aerial imaging using autonomous runway-free small flying wing vehicles. In *Proceedings of the IEEE International Conference on Geoscience and Remote Sensing*, pages 506–509, July 2008.

32. Gumstix. Gumstix verdex specifications, 2008. http://www.gumstix.com.

33. M. Baumann. Imager development and image processing for small UAV-based real-time multispectral remote sensing. Master's thesis, University of Applied Sciences Ravensburg-Weingarten and Utah State University, 2007.

34. Y. Han, A. M. Jensen, and H. Dou. Programmable multispectral imager development as light weight payload for low cost fixed wing unmanned aerial vehicles. In *Proceedings of the ASME Design Engineering Technical Conference Computers and Information in Engineering*, number MESA-87741, September 2009.

35. gPhoto, 2008. http://www.gphoto.org/.

36. Picture Transfer Protocol, 2008. http://en.wikipedia.org/wiki/Picture_Transfer_Protocol.

37. P. Brisset, A. Drouin, M. Gorraz, P. S. Huard, and J. Tyler. The Paparazzi solution. In *Micro Aerial Vehicle (MAV) 2006*, November 2006.

Chapter 3

Attitude Estimation Using Low-Cost IMUs for Small Unmanned Aerial Vehicles

An inertial measurement unit (IMU) is a device to measure the relative states of a static or mobile unit with respect to the inertial reference frame. Recently, many microelectromechanical systems (MEMS) IMUs have emerged for only several hundred U.S. dollars [1]. These low-cost IMUs can be used on unmanned vehicles for navigation [2], or can be combined with imaging sensors for georeferencing purposes. For example, the accurate orientation data are needed for the interpretation of the images from an airborne LIDAR radar. Actually, an accurate IMU accounts for a large portion of the total cost for an unmanned autonomous system [3]. The emergence of low-cost IMUs makes it possible to use more unmanned vehicles for agricultural or environmental applications such as precision farming and real-time irrigation control [4,5]. With the current trend of modularization and standardization in the unmanned system design, the developers can either use a cheap commercial-off-the-shelf (COTS) IMU as a part of the navigation system or develop their own system with low-cost inertial sensors.

In this book, the low-cost IMUs are defined as those with the price around or less than $3000. Low-cost MEMS IMUs are widely used on small or micro unmanned vehicles since they are small, light, yet still powerful. However, these cheap IMUs have bigger measurement errors or noise compared with expensive navigation grade or tactical grade IMUs [6]. It is challenging to design, test, and integrate these low-cost inertial sensors into a powerful IMU for navigation uses. More considerations for system design and sensor fusion algorithms need to be addressed to achieve autonomous navigation missions.

IMUs are usually used to measure the vehicle states such as orientation, velocity, and position. The orientation measurement is especially important for missions

Remote Sensing and Actuation Using Unmanned Vehicles, First Edition. Haiyang Chao and YangQuan Chen.
© 2012 by The Institute of Electrical and Electronics Engineers, Inc.
Published 2012 by John Wiley & Sons, Inc.

requiring accurate navigation. However, the orientation is not directly measurable with the current COTS MEMS sensors. It has to be estimated from a set of correlated states such as angular rates (gyros), linear accelerations (accelerometers), and magnetic fields (magentometers). Therefore, the estimation accuracy of IMUs heavily relies on the sensor fusion algorithm. Many researchers have looked into the state estimation problem using nonlinear filtering techniques [7]. Different kinds of Kalman filters are widely used in the aeronautics societies for spacecraft attitude estimations [8]. However, many of these algorithms are developed for highly accurate inertial sensors. Besides, those algorithms may have high demands for the computational power, which may not be possible for low-cost IMUs. The extended Kalman filter is introduced for IMUs with gyros, accelerometers, and magnetometers such as MNAV from Crossbow Technology as part of an open-source project [9]. However, this method assumes the UAV acceleration to be zero for the measurement equation, which may cause errors in the pitch estimation especially while turning. The measurement of the air speed or the ground speed can be incorporated into the filter for more accurate estimation of the acceleration on each body axis [10]. There are also other approaches to this problem, like complementary filters [11]. With the current trend of modularization and standardization in the UAV design, the UAV developers can save a large amount of time by buying the cheap commercial-off-the-shelf (COTS) IMUs and configure them into a complete navigation system. The systematic procedure for the state filter design, tuning, and validation needs to be developed to support these cheap COTS IMUs.

This chapter focuses on the implementation of the state estimation filter together with in-flight filter testing and comparison for small UAV applications. First, the problem definition and preliminaries of the rigid body rotations are discussed. Then, a comparative study on low-cost IMU hardware and sensor suites is provided with several example COTS IMUs. The attitude estimation software is further focused with both complementary filter approaches and EKF approaches. Finally, experimental results from real flight tests show the estimation results from typical estimation algorithms.

3.1 STATE ESTIMATION PROBLEM DEFINITION

The state estimation problem is defined in this section in consideration of small UAV constraints. The commonly used UAV states include the following:

- *Position*: For example longitude (p_e), latitude (p_n), altitude (h) (LLH);
- *Velocity*: Defined in the body frame or Earth frame, (u), (v), (w);
- *Attitude*: The representation can be Euler angles: roll (ϕ), pitch (θ), and yaw (ψ), or any other format such as quaternion or rotation matrix;
- *Rate gyro*: Angular velocity expressed in the body frame (p, q, r);
- *Acceleration*: Acceleration expressed in the body frame (a_x, a_y, a_z);
- Air speed (v_a), angle of attack (α), and slide-slip angle (β).

The following states need extra estimation filters since no direct measurements are available or the update rate is not fast enough:

- *Position*: The position information can greatly affect the georeferencing result. However, many civilian GPS receivers can only provide measurements at 4 Hz with the 3D accuracy of more than 3 m.

- *Attitude*: The orientation information is very important for both flight control and image georeferencing.

- *Wind Speed*: The wind, especially the wind gust, is a key factor for UAV flight control and the weather prediction.

To make an accurate estimation of the above system states, the UAV motions need to be characterized first. Here, the UAVs are assumed to be rigid bodies moving in Euclidean space. The rigid body rotations are reviewed first and then the UAV kinematics equation is formulated.

3.2 RIGID BODY ROTATIONS BASICS

Rigid body rotations are to model the rotation movements of rigid bodies between the body frame and the inertial reference frame. There are several representations available including Euler angles, rotation matrix, and unit quaternion. Euler angles are the most intuitive way to model the rotations. However, Euler angles have gimbal lock problems at certain extreme orientations [12]. Rotational matrix and unit quaternion do not have the gimbal lock problem through, with more variables used. In this section, different representations of rotations are introduced together with conversions between each other. Then, the kinematic equation is included with different formats to form a basis for the attitude estimation analysis.

3.2.1 Frame Definition

To describe the vehicle movements in 3D space, the coordinate frames are defined as follows, shown in Fig. 3.1:

(1) *Vehicle Body Frame*: F_{body}, the reference frame with the origin at the UAV gravity center and the xyz axes pointing forward, right and down.

(2) *Inertial Navigation Frame*: F_{nav}, the reference frame with a specific ground origin and the axes usually pointing the North, East and down to the Earth center. The user can also define their specific navigation frame.

(3) *Earth-Centered Earth-Fixed (ECEF) Frame*: F_{ECEF}, the reference frame with the origin at the Earth center. The z-axis passes through the north pole, the x-axis passes through the equator at the prime meridian, and the y-axis passes through the equator at $90°$ longitude.

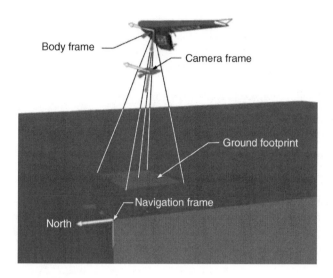

Figure 3.1   Aircraft coordinate systems [5].

3.2.2   Rotation Representations

Euler Angles   Euler angles are the most intuitive way to model the rotations, with only three variables, roll (ϕ), pitch (θ), and yaw (ψ). Given the reference frame and the body frame, Euler angles could be calculated through three basic rotations (each around one axis). The most common notation is the yaw-pitch-roll rotation or so-called $z - y - x$ rotation [13]. The rotation sequence from the navigation frame to the body frame with the angle limit is specified as:

- right-handed rotation $\psi \in (-180, 180]$ about the z-axis by the yaw angle.
- right-handed rotation $\theta \in [-90, 90]$ about the new y-axis by the pitch angle.
- right-handed rotation $\phi \in (-180, 180]$ about the new x-axis by the roll angle.

Rotation Matrix   Rotation matrix is the matrix that transforms a vector expressed in the body frame to the corresponding one expressed in the reference frame [14]. It is defined as follows:

$$R = \begin{bmatrix} r_{11} & r_{12} & r_{13} \\ r_{21} & r_{22} & r_{23} \\ r_{31} & r_{32} & r_{33} \end{bmatrix} \quad s.t. \quad RR^T = I, \quad \det(R) = 1, \tag{3.1}$$

where r_{ij} is a scalar, R^T is the transpose of matrix R.

The space of rotation matrices is also denoted as SO(3) (Special Orthogonal Group), which are also called direction cosine matrices. It is worth mentioning here that the column vectors of R are actually the unit vectors of the body frame expressed in the inertial reference frame and the row vectors of R are actually the unit vectors of the inertial reference frame expressed in the body frame [2]. For example, the vector $[1, 0, 0]^T$ expressed in the body frame is the same with the vector $[r_{11}, r_{21}, r_{31}]^T$ expressed in the inertial reference frame because $R[1, 0, 0]^T = [r_{11}, r_{21}, r_{31}]^T$.

Unit Quaternion   Unit quaternion is another way of accurate modeling of rotations using only four variables [14]. It is defined as follows:

$$q = \begin{bmatrix} q_0 \\ q_1 \\ q_2 \\ q_3 \end{bmatrix}, \quad s.t. \quad |q|^2 = q_0^2 + q_1^2 + q_2^2 + q_3^2 = 1. \tag{3.2}$$

The unit quaternion is usually denoted as $(q_0, \vec{q})$ with $\vec{q} = [q_1, q_2, q_3]^T$.

3.2.3   Conversion Between Rotation Representations

These three representations can all be used to model the rotation of the UAV. Euler angle is widely used in the guidance, navigation, and controls because of its simplicity. However, Euler angles have the singularity problem at some extreme points, which is also called gimbal lock problem. The rotation matrix or quaternion format is more widely used in the attitude estimation problem due to their completeness and constraints. For instance, the rotation matrix must maintain its orthogonality among vectors. Thus, the conversion between any two of them is also needed for the estimation and control problems of small UAVs.

Euler Angles to Rotation Matrix   This conversion is needed to make a vector mapping from the body frame to the reference inertial frame.

$$R = \begin{bmatrix} c\theta c\psi & -c\phi s\psi + s\phi s\theta c\psi & s\phi s\psi + c\phi s\theta c\psi \\ c\theta s\psi & c\phi c\psi + s\phi s\theta s\psi & -s\phi c\psi + c\phi s\theta s\psi \\ -s\theta & s\phi c\theta & c\phi c\theta \end{bmatrix}, \tag{3.3}$$

where c represents cos, s represents sin.

Rotation Matrix to Euler Angles   This conversion is needed if the rotation matrix is used in the estimation equation while Euler angles are needed for other purposes such as controls.

$$\phi = \tan^{-1}\left(\frac{r_{32}}{r_{33}}\right), \tag{3.4}$$

$$\theta = -\sin^{-1}(r_{31}), \tag{3.5}$$

$$\psi = \tan^{-1}\left(\frac{r_{21}}{r_{11}}\right), \tag{3.6}$$

where r_{ij} is from the rotation matrix defined in (3.1).

Quaternion to Euler Angles   This conversion is needed if the rotation matrix is chosen in the estimation problem and Euler angles are needed.

$$\phi = \tan^{-1}\frac{2(q_2q_3 + q_0q_1)}{q_0^2 - q_1^2 - q_2^2 + q_3^2}, \tag{3.7}$$

$$\theta = \sin^{-1}(-2(q_1q_3 - q_0q_2)), \tag{3.8}$$

$$\psi = \tan^{-1}\frac{2(q_1q_2 + q_3q_0)}{q_0^2 + q_1^2 - q_2^2 - q_3^2}, \tag{3.9}$$

where $[q_0, q_1, q_2, q_3]^T$ is the unit quaternion.

Quaternion to Rotation Matrix

$$R = \begin{bmatrix} 1 - 2(q_2^2 + q_3^2) & 2(q_1q_2 - q_0q_3) & 2(q_0q_2 + q_1q_3) \\ 2(q_1q_2 + q_0q_3) & 1 - 2(q_1^2 + q_3^2) & 2(q_2q_3 - q_0q_1) \\ 2(q_1q_3 - q_0q_2) & 2(q_0q_1 + q_2q_3) & 1 - 2(q_1^2 + q_2^2) \end{bmatrix}, \tag{3.10}$$

where $[q_0, q_1, q_2, q_3]^T$ is the unit quaternion.

3.2.4   UAV Kinematics

Kinematics is to describe the motion of the UAV without considering the forces and torques that cause it [12]. UAV kinematics can be used for estimation of the UAV orientation without knowing the accurate dynamic model.

3.2.4.1   Angle Kinematic Equation

The angle kinematic equation deals with the derivatives of the orientations. They can be expressed in Euler angles, unit quaternion, or rotation matrix.

Euler Angles Representation [12]

$$\dot{\phi} = p + q \sin \phi \tan \theta + r \cos \phi \tan \theta, \tag{3.11}$$

$$\dot{\theta} = q \cos \phi - r \sin \theta, \tag{3.12}$$

$$\dot{\psi} = \frac{q \sin \phi + r \cos \phi}{\cos \theta}, \tag{3.13}$$

where p, q, r are the angular rates expressed in the body frame.

Quaternion Representation [12]

$$\dot{q} = \frac{1}{2} \begin{bmatrix} 0 & -p & -q & -r \\ p & 0 & r & -q \\ q & -r & 0 & p \\ r & q & -p & 0 \end{bmatrix} q. \tag{3.14}$$

Rotation Matrix Representation

$$\dot{R} = R\Omega_{\times} = (R\Omega)_{\times} R, \quad \Omega = \begin{bmatrix} p \\ q \\ r \end{bmatrix}, \quad \Omega_{\times} = \begin{bmatrix} 0 & -r & q \\ r & 0 & -p \\ -q & p & 0 \end{bmatrix}, \tag{3.15}$$

where R is the rotation matrix from the body frame to the earth inertial frame. R is also denoted as the attitude of the body-fixed frame relative to the earth inertial frame.

The above equations can be used for the estimation of the UAV orientation given the angular rate measurements. However, there must be some compensations because MEMS gyro sensors can have big drifts and random noises.

3.2.4.2 Acceleration Kinematic Equation

Acceleration kinematic equation describes how to get the acceleration measurements with velocity derivatives. It can be used to correct the attitude estimation.

$$\begin{bmatrix} \dot{v}_x \\ \dot{v}_y \\ \dot{v}_z - g \end{bmatrix} = R_{vb} \begin{bmatrix} a_x \\ a_y \\ a_z \end{bmatrix}, \tag{3.16}$$

where v_x, v_y, v_z are GPS velocity measurements in the navigation frame, R_{vb} is the rotation matrix from the body frame to the vehicle navigation frame, and a_x, a_y, a_z are the accelerometer readings measured in the body frame.

3.3 LOW-COST INERTIAL MEASUREMENT UNITS: HARDWARE AND SENSOR SUITES

In the past decade, GPS and inertial sensors have become increasingly smaller, lighter, and cheaper, which can be easily fit on small or micro UAVs. These inertial sensors (gyros, accelerometers, and others) can be combined into an IMU to make three-dimensional estimations. In this section, the IMU basic and sensor hardware are introduced with simple attitude estimation algorithms relying only on one sensor.

3.3.1 IMU Basics and Notations

The basic function of an IMU is to measure the movements of a craft or a vehicle in 3D space. An IMU usually comprises a sensor pack, an embedded processor, and its supporting software. The system structure of an IMU is shown in Fig. 3.2. The sensor pack can include three-axis gyro/accelerometer/magnetometer, GPS, pressure, or vision sensors. The software includes sensor calibration function and state estimation filters.

The available direct measurements for low-cost IMUs and GPS include the following:

(1) *Position*: for example longitude (p_e), latitude (p_n), altitude (h) (LLH) from GPS in 4–10 Hz mostly, the altitude or height can also be measured by pressure or ultrasonic sensors;

(2) *Velocity*: ground speed from GPS (v_n, v_e, v_d) and the air speed from pressure sensors;

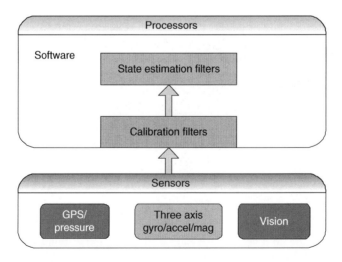

Figure 3.2   System structure for IMUs.

(3) *Rate Gyro*: angular velocity expressed in the body frame (p, q, r);

(4) *Acceleration*: linear acceleration expressed in the body frame (a_x, a_y, a_z).

However, some key states are not directly measurable such as the attitude. IMUs rely on the sensor fusion algorithm to provide an accurate estimation of attitudes and positions. The sensor fusion problem is defined as making an optimal estimation of the required vehicle states with the direct measurements from multiple sensors. This problem is also called a state estimation or a nonlinear filtering problem [7]. There are many possible solutions to this problem such as Kalman filters or complementary filters.

3.3.2 Sensor Packs

The developments of the low-cost MEMS inertial sensors can be traced back as early as 1970s [6]. In this section, the possible sensor packages for IMUs are introduced with an emphasis on the error models and the IMU categories.

Gyro   A gyro sensor is to measure the angular rate around the prespecified axis observed from the earth coordinated in the body frame. Most manned or unmanned aircraft have three-axis gyros onboard. The gyro error model can be expressed as

$$\hat{\omega} = (1 + s_g)\omega + b_g + \mu_g, \tag{3.17}$$

where $\hat{\omega}$ is the measurement value, s_g is the scale error, ω is the true value, b_g is the gyro bias, and μ_g is the random noise. Gyro sensors can be integrated to get the estimate of the angle. However, angle estimates based only on gyros may have big drifts since the MEMS gyro bias is integrated over the time. The angle estimation results using only MEMS gyro sensors are shown in Fig. 3.3. The drift can be clearly observed. The true value is provided by a commercial IMU (Microstrain GX2). The complex gyro algorithm is described in Eqn. (1.15) and Eqn. (1.16). The simple gyro

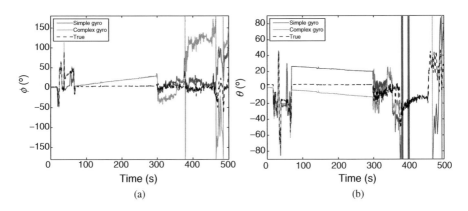

Figure 3.3   Attitude estimation using gyro only. (a) Roll estimation. (b) Pitch estimation.

algorithm can be expressed as follows:

$$\phi = \int p \ dt, \tag{3.18}$$

$$\theta = \int q \ dt. \tag{3.19}$$

Accelerometer   Accelerometers used on low-cost IMUs are to measure the linear acceleration. In fact, accelerometers measure the acceleration minus the gravity vector. For example, the default output of the accelerometer (static) is -1 when the axis is pointing down into the earth center.

$$\hat{a} = (1 + s_a)a + b_a + \mu_a, \tag{3.20}$$

where $\hat{a}$ is the measurement value, s_a is the scale error, a is the true value, b_a is the accelerometer bias, and μ_a is the random noise.

The accelerometer can also be used to measure the vehicle attitude since three-axis accelerometers can measure the gravity vector under the condition of zero acceleration. However, angle estimates from accelerometers suffer from high-frequency noise when the unmanned vehicles are moving. The roll and pitch estimation using accelerometers as tilt sensors can be expressed as follows:

$$\phi = \tan^{-1} \frac{a_y}{a_z}, \tag{3.21}$$

$$\theta = \sin^{-1} a_x, \tag{3.22}$$

where $a_x, a_y,$ and a_z are in G, and ϕ and θ are in radians, all expressed in the body frame of the unmanned vehicle. The roll and pitch estimation relying only on accelerometers are shown in Fig. 3.4. The high-frequency noise is from the vibration of the aircraft.

Magnetometer   Magnetometers are to measure the magnetic fields of the Earth, which can be approximated as an absolute value assuming the vehicle is not moving

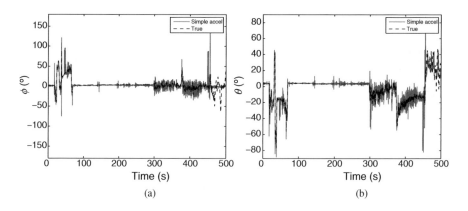

Figure 3.4   Attitude estimation using accelerometer only. (a) Roll estimation. (b) Pitch estimation.

too fast. Three-axis magnetometer can be used for heading estimation and gyro bias compensation. One disadvantage of magnetometer sensors is that the hard-iron and soft-iron calibrations are needed for every vehicle.

GPS GPS sensors can provide measurements of the absolute position, velocity, and course angle. The position packets can either be latitude, longitude, height (LLH) or x, y, z expressed in the ECEF frame. The velocities include v_n, v_e, v_d, all with respect to the inertial frame. The GPS measurements have advantages of bounded errors, which can be used to reset the system error infrequently. The disadvantages of GPS include low update rate (<4 Hz for most low-cost GPS) and vulnerability to weather and terrain interference. The course angle is defined as the angle relative to the north clockwise [15]. It can be calculated using the measured GPS velocity, as shown in below:

$$course = \tan^{-1}\frac{v_e}{v_n}, \tag{3.23}$$

The disadvantage of using GPS for course estimation is the speed requirement.

3.3.3 IMU Categories

Based on the performance and the characteristics of the above sensors, the commercial IMUs can be categorized into four types: navigation grade, tactical grade, industrial grade, and hobbyist grade. It is worth mentioning here that most of the industrial grade and hobbyist grade IMUs use MEMS inertial on-chip sensors, which greatly reduce the unit sizes and weights. The brief specifications are shown in Table 3.1. It can be seen that low-cost IMUs mostly fall into the industrial grade or the hobbyist grade due to their low cost and bigger errors compared with navigation or tactical grade IMUs.

3.3.4 Example Low-Cost IMUs

Several example low-cost IMUs are introduced in this section with detailed specifications. There are two types of low-cost IMUs available, with or without orientation estimates. Some IMUs are even coupled with GPS receivers for better accuracy.

Table 3.1 IMU Categories

IMU Type	Navigation Grade	Tactical Grade	Industrial Grade	Hobbyist Grade
Cost ($)	$>$50k	10–20k	0.5–3k	$<$500
Weight	$>$5 lb	About 1 lb	$<$5 oz	
Gyro bias	$<$0.1 deg/h	0.1–10 deg/h	$\leq$1 deg/s	$>$1 deg/s
Gyro random Walk error	$<$0.005 deg/$\sqrt{h}$	0.2–0.5 deg/$\sqrt{h}$		
Accel bias	5–10 μg	0.02–0.04 mg		
Example	Honeywell HG9848	Honeywell HG1900	Microstrain GX2	ArduIMU

(a)   (b)

Figure 3.5 Example COTS IMUs. (a) Microstrain GX2. (b) Xsens Mti-g IMU [16].

A typical COTS IMU, Microstrain GX2 is shown in Fig. 3.5a, which has tri-axis gyros, accelerometers, and magnetometers. Microstrain GX2 supports USB, RS232, or RS422 connections. For comparison reasons, Xsens Mti-g, a relatively more expensive IMU ($5000–$10000), is also shown in Fig. 3.5b. Xsens Mti-g has a GPS receiver integrated. Their specifications are shown in the later table.

In recent years, even cheaper IMUs (several hundred U.S. dollars) become available due to the wide use of inertial sensors on consumer electronics. These kind of IMUs usually do not come with the orientation estimation firmware. One representing example is the Ardu IMU flat version, which uses flat mounted gyro and accelerometer chips, shown in Fig. 3.6a. Traditional IMUs have sensors mounted vertically (e.g., ADIS 16405 IMU), shown in Fig. 3.6b as part of the AggieNav [17].

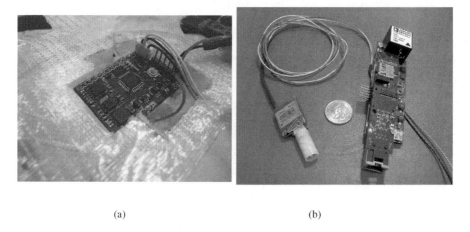

(a)   (b)

Figure 3.6 Low-cost IMUs (no orientation firmware). (a) Ardu-IMU. (b) AggieNav using ADIS IMU [5].

Table 3.2 IMU Specification Comparisons

Specifications	Microstrain 3DM-GX2	Xsens Mti-g (w. GPS)	ADIS16405	Ardu-IMU
Size (mm)	$41 \times 63 \times 32$	$58 \times 58 \times 33$	$23 \times 23 \times 23$	$28 \times 39 \times 12$
Weight (g)	39	68	16	6
Orientation accuracy				
(static)	$\pm 0.5^{\text{deg}}$ typical	$<0.5^{\text{deg}}(\phi/\theta)$	Only raw data	N/A
(dynamic)	$\pm 2.0^{\text{deg}}$ typical	1^{deg} RMS		
Update rate	$<200\,\text{Hz}$	$<120\,\text{Hz}$	$<330\,\text{Hz}$	$<50\,\text{Hz}$
	w. raw sensor data			

The key specifications from the above IMUs are compared in Table 3.2. Actually, it is still challenging to achieve degree level accuracy with the current low-cost IMU hardware under dynamic environments.

3.4 ATTITUDE ESTIMATION USING COMPLEMENTARY FILTERS ON SO(3)

Given the low-cost sensor pack, the next step will be to design an attitude estimation algorithm for the vehicle. Complementary filters have been applied on the attitude estimation problem for many years due to their simple representation and advantages in combining estimates from different sensors. In recent years, nonlinear complementary filters on special orthogonal groups SO(3) have been proposed for better estimation accuracy [11]. Such filters are especially suited for implementation on embedded processors because its low computation demands.

SO(3) is the special orthogonal group of 3×3 matrices, which satisfies the following property:

$$SO(3) = \{R \in \mathbb{R}^{3 \times 3} : RR^T = I, \quad detR = +1\}.$$

Matrices in SO(3) can be used to represent the rigid body orientation in 3D space. The angle kinematics equation is

$$\dot{R} = R\Omega_\times = R \begin{bmatrix} 0 & -r & q \\ r & 0 & -p \\ -q & p & 0 \end{bmatrix}, \tag{3.24}$$

The above equation can be integrated to get the orientation assuming no gyro drift:

$$R_{t+1} = R_t \begin{bmatrix} 1 & -r_t \Delta t & q_t \Delta t \\ r_t \Delta t & 0 & -p_t \Delta t \\ -q_t \Delta t & p_t \Delta t & 0 \end{bmatrix}, \tag{3.25}$$

where Δt is the time step for integration.

Instead of using gyro directly, the nonlinear complementary filters on SO(3) [11] add correction terms to compensate for the gyro drifts by comparing with estimates from other sensors such as accelerometers or magnetometers. A general form of complementary filters on SO(3) can be expressed as

$$\dot{\hat{R}} = P(\Omega) + C(\hat{R}, a, m), \tag{3.26}$$

where $P(\Omega)$ is the prediction term dominated by the gyro readings Ω, a is the acceleration vector, m is the magnetic vector, and $C(\hat{R}, a, m)$ is the correction term derived from the estimation error using other sensor sources. Several nonlinear complementary filters on SO(3) have been introduced with different prediction terms ($P(\Omega)$) and correction terms ($C(\hat{R}, a, m)$), including direct complementary filter, passive complementary filter, and explicit complementary filter [11]. Two representing filters are introduced in detail in below.

3.4.1   Passive Complementary Filter

The basic idea of passive complementary filter is to use acceleration/magnetic estimates to correct the gyro updates [11]. The correction scheme can either be a basic proportional scheme or a proportional plus integral scheme. A basic passive complementary filter with proportional correction is described as below [11]:

$$\dot{\hat{R}} = \hat{R}(\Omega_\times + k_P \mathcal{P}_a(\hat{R}^T R_y)), \tag{3.27}$$

$$\mathcal{P}_a(\tilde{R}) = \frac{1}{2}(\hat{R}^T R_y - R_y^T \hat{R}), \tag{3.28}$$

where k_P is the gain to be tuned, R_y is the rotation matrix estimated with acceleration and/or magnetic sensors.

3.4.2   Explicit Complementary Filter

Explicit complementary filter is another complementary filter with much simpler representations [11]. A basic explicit complementary filter with proportional correction can be formulated as below [11]:

$$\dot{\hat{R}} = \hat{R}(\Omega_\times + k_P(\omega_{else})_\times), \tag{3.29}$$

$$\omega_{else} = \sum_{i=1}^{n} k_i v_i \times \hat{v}_i, \quad k_i > 0, \tag{3.30}$$

where k_P is the gain to be tuned, v_i is the body frame observation of the inertial direction, and $\hat{v}_i$ is an estimate of v_i using $\hat{R}$.

3.4.3 Flight Test Results

The passive complementary filter results are tested using the COTS Microstrain GX2 IMU. The raw sensor data including three-axis gyro, accelerometer, and magnetometer were logged in 100 Hz during flights. The raw data are then feed into the complementary filter in MATLAB. The attitude estimation from both passive complementary filter and Microstrain is shown in Fig. 3.7. The estimation error is shown in Fig. 3.8. It can be observed that the estimation accuracy of passive complementary

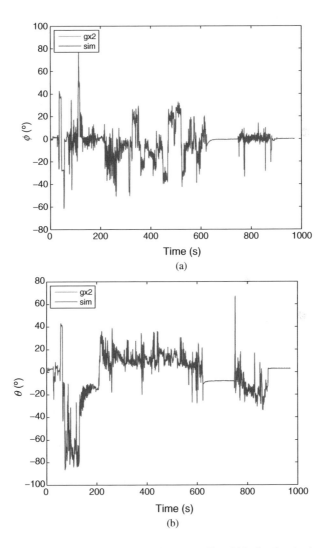

Figure 3.7   Attitude estimation results with complementary filter. (a) Roll estimation between CF (sim) and GX2. (b) Pitch estimation between CF (sim) and GX2.

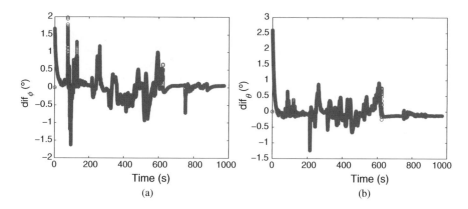

Figure 3.8   Estimation error comparison (complementary filter vs. GX2). (a) Roll error. (b) Pitch error.

is within two degrees most of the time and a large portion of the error comes from the initialization.

3.5 ATTITUDE ESTIMATION USING EXTENDED KALMAN FILTERS

Extended Kalman filters are frequently used in nonlinear estimation problems, especially the attitude estimation problem of rigid bodies such as a spacecraft or an aircraft. The extended Kalman filter can recursively estimate the system states from system measurements corrupted with Gaussian noises. It has advantages here since both gyro and accelerometer sensors have drifts and Gaussian-like noises. The general extended Kalman filter and two representing approaches for the attitude estimation problem are introduced in the following sections.

3.5.1 General Extended Kalman Filter

Assume that a general nonlinear discrete-time system can be modeled as follows:

$$x_k = f(x_{k-1}, u_k) + w_k, \quad s.t. \quad w \sim N(0, Q), \tag{3.31}$$

$$y_k = g(x_k) + v_k, \quad s.t. \quad v \sim N(0, R), \tag{3.32}$$

where x_k is a m×1 vector, y_k is a n×1 vector, $f(x_{k-1}, u_k)$, and $g(x_k)$ are nonlinear functions. The first equation is called propagation equation and the second one is called the measurement equation.

Given the initial values P_0, the measurement covariance R and the state covariance Q, the optimal Kalman estimate of the states can be updated using the following steps [18]:

(1) State estimation extrapolation: $\hat{x}_{k|k-1} = f(x_{k-\hat{1}|k-1}, u_k)$,

(2) Error covariance extrapolation: $P_{k|k-1} = F_k P_{k-1|k-1} F_k^T + Q_k$,

(3) Kalman gain: $K_k = P_{k|k-1}G_k^T(G_k P_{k|k-1}G_k^T + R_k)^{-1}$,

(4) State estimate update: $\hat{x}_{k|k} = x_k|\hat{k}-1 + K_k(y_k - g(x_k|\hat{k}-1))$,

(5) Error covariance update: $P_{k|k} = (I - K_k G_k)P_{k|k-1}$.

The F_k is the Jacobian matrix of $f(x_{k-1}, u_k)$, and the G_k is the Jacobian matrix of $g(x_k)$.

3.5.2   Quaternion-Based Extended Kalman Filter

Unit quaternion has many applications in the state estimation problems because of its simplicity. A quaternion-based extended Kalman filter is proposed originally for the MNAV IMU from Crossbow Technology [9]. The system state variables include both the unit quaternion q and the gyro biases. The measurements or observation of the system is the acceleration: a_x, a_y, a_z, and the yaw angle ψ derived from the magnetometer. The propagation equation and the measurement equation are listed below [9].

$$\dot{q} = \frac{1}{2}\begin{bmatrix} 0 & -\hat{p} & -\hat{q} & -\hat{r} & 0 & 0 & 0 \\ \hat{p} & 0 & \hat{r} & -\hat{q} & 0 & 0 & 0 \\ \hat{q} & -\hat{r} & 0 & \hat{p} & 0 & 0 & 0 \\ \hat{r} & \hat{q} & -\hat{p} & 0 & 0 & 0 & 0 \\ \hline 0 & 0 & 0 & 0 & 0 & 0 & 0 \\ 0 & 0 & 0 & 0 & 0 & 0 & 0 \\ 0 & 0 & 0 & 0 & 0 & 0 & 0 \end{bmatrix} q + w_k, \quad q = \begin{bmatrix} q_0 \\ q_1 \\ q_2 \\ q_3 \\ b_p \\ b_q \\ b_r \end{bmatrix}, \quad (3.33)$$

$$\begin{bmatrix} \hat{p} \\ \hat{q} \\ \hat{r} \end{bmatrix} = \begin{bmatrix} p \\ q \\ r \end{bmatrix} - \begin{bmatrix} b_p \\ b_q \\ b_r \end{bmatrix}, \quad (3.34)$$

$$\begin{bmatrix} a_x \\ a_y \\ a_z \\ \psi \end{bmatrix} = \begin{bmatrix} 2g(q_1 q_3 - q_0 q_2) \\ 2g(q_2 q_3 + q_0 q_1) \\ g(q_0^2 - q_1^2 - q_2^2 + q_3^2) \\ \tan^{-1}\frac{2(q_1 q_2 + q_3 q_0)}{q_0^2 + q_1^2 - q_2^2 - q_3^2} \end{bmatrix} + v_k \quad v_k = N(0, R). \quad (3.35)$$

It is worth pointing out that the measurement equation has an assumption that the acceleration measured is only the projection of the gravity vector. However, this assumption may not be true for small UAVs.

3.5.3   Euler Angles-Based Extended Kalman Filter

Euler angles can also be chosen as the system states for the attitude estimation problem. Assume that the system state is a vector x, representing the roll and pitch angle, and the system output is a vector $\hat{y}$, representing the accelerometer readings, the system can then be modeled [19]:

$$x = \begin{bmatrix} \phi \\ \theta \end{bmatrix}, \qquad \hat{y} = \begin{bmatrix} a_x \\ a_y \\ a_z \end{bmatrix}, \tag{3.36}$$

$$\dot{x} = \begin{bmatrix} p + q\sin\phi\tan\theta + r\cos\phi\tan\theta \\ q\cos\phi - r\sin\phi \end{bmatrix} + v_w, \quad v_w \sim N(0, Q), \tag{3.37}$$

$$\hat{y} = \begin{bmatrix} \dot{u} - rv + qw + g\sin\theta \\ \dot{v} + ru - pw - g\cos\theta\sin\phi \\ \dot{w} - qu + pv - g\cos\theta\cos\phi \end{bmatrix} + v_k, \quad v_k \sim N(0, R). \tag{3.38}$$

In fact, the velocities (u, v, w) are not easily measurable at a high frequency. The air speed can be used instead to simplify the measurement equation for small UAVs. The following assumptions can be made [19]:

- $\dot{u} = \dot{v} = \dot{w} = 0$. The small UAV will not accelerate all the time;
- $v = 0$. The small UAV will not go sideways;
- $u = v_a\cos\theta$, $w = v_a\sin\theta$;

where V_a is the air speed measured by the pitot tube. The measurement equation can then be simplified.

$$\hat{y} = \begin{bmatrix} qV_a\sin\theta + g\sin\theta \\ rV_a\cos\theta - pV_a\sin\theta - g\cos\theta\sin\phi \\ -qV_a\cos\theta - g\cos\theta\cos\phi \end{bmatrix} + v_k, v_k \sim N(0, R). \tag{3.39}$$

The attitude can then be estimated using the extended Kalman filter following the steps described in the above section.

3.6 AGGIEEKF: GPS-AIDED EXTENDED KALMAN FILTER

AggieEKF, a GPS-aided extended Kalman filter is proposed in this section with considerations from both filters designed in the above sections. Unit quaternion is chosen for our approach because of its simplicity and constrains from its unit scalar. An extended Kalman filter similar to the one developed by Jung and Liccardo [9] is developed. However, the measurement equation is replaced by a more accurate estimation of the gravity vector with the help from the GPS speed measurements. The system equations are shown as below, where V_g is the ground speed measured by the

GPS. The attitude state estimation can be calculated using the steps described in the above section.

$$\dot{q} = \frac{1}{2} \left[\begin{array}{ccc|cccc} 0 & -\hat{p} & -\hat{q} & -\hat{r} & 0 & 0 & 0 \\ \hat{p} & 0 & \hat{r} & -\hat{q} & 0 & 0 & 0 \\ \hat{q} & -\hat{r} & 0 & \hat{p} & 0 & 0 & 0 \\ \hat{r} & \hat{q} & -\hat{p} & 0 & 0 & 0 & 0 \\ \hline 0 & 0 & 0 & 0 & 0 & 0 & 0 \\ 0 & 0 & 0 & 0 & 0 & 0 & 0 \\ 0 & 0 & 0 & 0 & 0 & 0 & 0 \end{array} \right] q + w_k, \tag{3.40}$$

$$\begin{bmatrix} \hat{a}_x \\ \hat{a}_y \\ \hat{a}_z \end{bmatrix} = \begin{bmatrix} 2g(q_1 q_3 - q_0 q_2) \\ 2g(q_2 q_3 + q_0 q_1) \\ g(q_0^2 - q_1^2 - q_2^2 + q_3^2) \end{bmatrix} + v_k, v_k = N(0, R), \tag{3.41}$$

$$\text{where } \begin{bmatrix} \hat{a}_x \\ \hat{a}_y \\ \hat{a}_z \end{bmatrix} = \begin{bmatrix} a_x \\ a_y - rV_g + g \sin \phi_{t-1} \\ a_z + qV_g + g \cos \phi_{t-1} \end{bmatrix}. \tag{3.42}$$

To validate the proposed AggieEKF algorithm, a AggieAir2 UAV is modified to fit in different types of CTOS IMUs for sensor data collection. Two flight tests were performed at the Cache Junction research farm on November 16 and 23 2010 for the comparison of three IMUs including Microstrain GX2, AggieNav, and Crossbow MNAV. The central bay layout of the three IMUs is shown in Fig. 3.9.

Figure 3.9   Central bay layout of three IMUs.

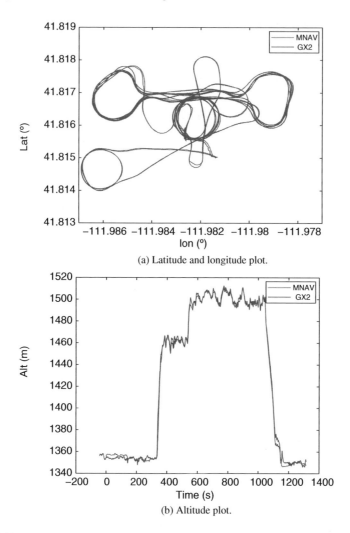

Figure 3.10   UAV trajectory for sensor data collection.

The AggieAir2 UAV uses Microstrain GX2 IMU and the Paparazzi TWOG autopilot as the onboard navigation control system. The UAV is commanded to follow the line 1–2 task, which comprises two circles (70 m radius) and a straight line connecting them. The UAV trajectory is plotted in Fig. 3.10. All the inertial sensor data are saved on the Gumstix while flying at 50 Hz, together with the timer on Gumstix. The GPS sensor data is also logged at 4 Hz. The AggieEKF is then implemented in the MATLAB for validations. The roll and pitch angle estimations from the AggieEKF algorithm are plotted together with the orientation outputs from the Microstrain GX2 IMU for reference in Figs. 3.11 and 3.12. It can be observed that the roll estimation is

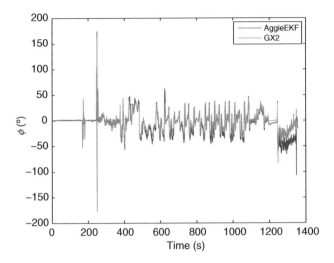

Figure 3.11 Roll angle estimation (AggieEKF).

good and the pitch estimation is a little bit noisy. As a comparison, the roll estimation based on the EKF algorithm developed by Jung and Liccardo [9] is also provided in Fig. 3.13. It is worth mentioning that it is hard to find a truth for the attitude estimation under dynamic cases for small UAVs because of the weight and space constraints.

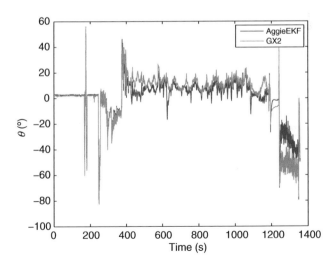

Figure 3.12 Pitch angle estimation (AggieEKF).

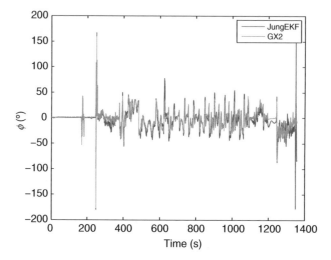

Figure 3.13   Roll angle estimation (EKF).

3.7   CHAPTER SUMMARY

In this chapter, the attitude estimation problem using low-cost IMUs is focused. The rigid body rotation basics and the IMU hardware are introduced first. Several estimation algorithms are compared including the complementary filter approach and the extended Kalman filter solution. The estimation results based from real flight data show the preliminary estimation results using both complementary and Kalman filters. The roll and pitch estimation from a complementary filter can achieve 2 degree accuracy compared with a commercial Microstrain GX2 IMU.

REFERENCES

1. J. Munoz and B. Premerlani. AuduIMU open source project, 2009. http://diydrones.com.

2. W. Premerlani. DCM estimation, 2009. http://gentlenav.googlecode.com.

3. H. Chao, Y. Cao, and Y. Q. Chen. Autopilots for small unmanned aerial vehicles: a survey. *International Journal of Control, Automation, and Systems*, 8(1):36–44, 2010.

4. H. Chao, A. M. Jensen, Y. Han, Y. Q. Chen, and M. McKee. AggieAir: towards low-cost cooperative multispectral remote sensing using small unmanned aircraft systems. In *Advances in Geoscience and Remote Sensing*, IN-TECH, Vukovar, Croatia, 2009.

5. H. Chao, M. Baumann, A. M. Jensen, Y. Q. Chen, Y. Cao, W. Ren, and M. McKee. Band-reconfigurable multi-UAV-based cooperative remote sensing for real-time water management and distributed irrigation control. In *Proceedings of the International Federal of Automatic Control (IFAC) World Congress*, pages 11744–11749, July 2008.

6. R. L. Greenspan. Inertial navigation technology from 1970-1995. *Journal of The Institute of Navigation*, 42(1):165–185, 1995.

7. J. L. Crassidis, J. L. Markley, and Y. Cheng. Nonlinear attitude filtering methods. *AIAA Journal of Guidance, Control, and Dynamics*, 30(1):12–28, 2007.

8. S.-G. Kim, J. L. Crassidis, Y. Cheng, and A. M. Fosbury. Kalman filtering for relative spacecraft attitude and position estimation. *AIAA Journal of Guidance, Control, and Dynamics*, 30(1):133–143, 2007.

9. J. S. Jang and D. Liccardo. Small UAV automation using MEMS. *IEEE Aerospace and Electronic Systems Magazine*, 22(5):30–34, 2007.

10. D. B. Kingston and R. W. Beard. Real-time attitude and position estimation for small UAVs using low-cost sensors. In *Proceedings of the AIAA 3rd Unmanned Unlimited Systems Conference and Workshop*, number AIAA-2007-6514, Chicago, IL, September 2004.

11. R. Mahony, T. Hamel, and J.-M. Pflimlin. Nonlinear complementary filters on the special orthogonal group. *IEEE Transactions on Automatic Control*, 53(5):1203–1218, 2008.

12. B. L. Stevens and F. L. Lewis. *Aircraft Control and Simulation*. 2nd ed., John Wiley & Sons, Hoboken, NJ, 2003.

13. W. F. Phillips. *Mechanics of Flight*. John Wiley & Sons, Hoboken, NJ, 2004.

14. R. M. Murray, Z. Li, and S. S. Sastry. *A Mathematical Introduction to Robotic Manipulation*. CRC Press, Boca Raton, FL, 1994.

15. u-blox company. u-blox GPS protocol, 2010. http://www.u-blox.com.

16. Xsens Company. Xsens-Mtig IMU specifications, 2008. http://www.xsens.com.

17. C. Coopmans. AggieNav: a small, well integrated navigation sensor system for small unmanned aerial vehicles. In *Proceedings of the ASME Design Engineering Technical Conference Computers and Information in Engineering*, number MESA-87636, September 2009.

18. G. Welch and G. Bishop. An Introduction to the Kalman Filter. Report, University of North Carolina at Chapel Hill, 2006. http://www.cs.unc.edu/ welch/kalman/kalmanIntro.html.

19. R. Beard, D. Kingston, M. Quigley, D. Snyder, R. Christiansen, W. Johnson, T. Mclain, and M. Goodrich. Autonomous vehicle technologies for small fixed wing UAVs. *Journal of Aerospace Computing, Information, and Communication*, 5(1):92–108, 2005.

Chapter 4

Lateral Channel Fractional Order Flight Controller Design for a Small UAV

4.1 INTRODUCTION

The lateral flight controller is to control the movement around the roll axis, which is critical to the flight stability. This chapter introduces a fractional order controller design for the roll-channel of a small fixed-wing UAV for better flight performance. Many researchers have looked into the problem of UAV modeling and control. Open-loop steady-state flight experiments are proposed for the aileron-(roll rate) and elevator-(pitch rate) loop system identification [1]. But the open-loop system identification has to have special requirements on UAV flight stability, which limits the roll and pitch reference signals to be as small as 0.02 rad. UAV model identification (ID) experiments can also be performed with human operators controlling the UAVs remotely. Different types of autoregressive with exogenous input (ARX) models are identified while the UAV is flying in loiter mode [2]. Human operators could generate open-loop responses, but it may be impossible for some specially designed reference such as pseudo random binary sequences (PRBS). Other researchers also tried closed-loop system identification method on separate channels of unmanned helicopters [3–5].

There are trade-offs, such as safety and maneuverability, while designing UAV system identification experiments. The system ID experiments are not easy to repeat since the UAV system could easily stall, given a too aggressive control input. On the other hand, very small excitations may not be adequate to excite the system dynamics. A closed-loop system identification method is used in this chapter with considerations for flight stability and test difficulty. The UAV is first roughly tuned with a set of initial proportional integral derivative (PID) parameters sufficient to guarantee stability while flying horizontally. Then the UAV initial closed-loop model

Remote Sensing and Actuation Using Unmanned Vehicles, First Edition. Haiyang Chao and YangQuan Chen.
© 2012 by The Institute of Electrical and Electronics Engineers, Inc.
Published 2012 by John Wiley & Sons, Inc.

is identified and the controllers are designed based on the identified models, which will be discussed in detail in this chapter.

Fractional order control (FOC) has attracted a lot of interest recently. FOC introduces new fractional derivative and fractional integral operators to the classical PID control. It provides additional design freedom for the controller tuning [6]. FOC has advantages in application scenarios such as servo control [7], water tank control, quad rotor [8], and other industrial applications [9]. The fractional order proportional integral (PI^λ) controller is one of the simplest fractional order controllers similar to the classical proportional integral (PI) controller. FOC can have advantages over traditional controllers because FOC introduces fractional order operators [10]. To simplify the flight control problem, the aileron-roll loop is singled out for controller comparisons between the fractional order PI (FOPI) controller and the integer order PID controller. The proposed controllers are tested in conditions such as strong wind gusts and payload variations in simulation and real-flight tests.

4.2 PRELIMINARIES OF UAV FLIGHT CONTROL

UAV dynamics can be modeled using system states including:

(1) *Position*: for example, longitude (p_e), latitude (p_n), height (h) (LLH)

(2) *Velocity*: three axis (u), (v), (w)

(3) *Attitude*: roll (ϕ), pitch (θ), and yaw (ψ)

(4) *Gyro Rate*: gyro acceleration p, q, r

(5) *Acceleration*: acceleration a_x, a_y, a_z

(6) Air speed (v_a), ground speed (v_g), angle of attack (α), and slide-slip angle (β).

UAV control inputs generally include aileron (δ_a), elevator (δ_e), rudder (δ_r), and throttle (δ_t). There are also elevons that combine the functions of the aileron and the elevator. Elevons are frequently used on flying-wing airplanes. Different types of UAVs may have different control surface combinations. For example, some delta-wing UAVs can just have elevator, aileron, and throttle with no rudder control.

The six degrees of freedom UAV dynamics can be modeled by a series of non-linear equations.

$$\dot{x} = f(x, u), \tag{4.1}$$

$$x = [p_n \ p_e \ h \ u \ v \ w \ \phi \ \theta \ \psi \ p \ q \ r]^T, \tag{4.2}$$

$$u = [\delta_a \ \delta_e \ \delta_r \ \delta_t]^T. \tag{4.3}$$

The ultimate objective of UAV flight control is to let the UAV follow a preplanned 3D trajectory with prespecified orientations. Due to the limits from the hardware, most current UAV autopilots can only achieve the autonomous waypoints navigation objective. There are basically two types of controller design approaches: the precise-model-based nonlinear controller design and the in-flight-tuning-based PID controller design. The first method requires a precise and complete dynamic model, which is usually very expensive to obtain. On the other hand, it is estimated that more than 90% of the current working controllers are PID controllers [11]. Most commercial UAV autopilots use cascaded PID controllers for autonomous flight control [12].

The cascaded PID controller can be used for UAV flight control because the non-linear dynamic model can be linearized around certain trimming points and be treated as a simple single-input and single-output (SISO) or multiple-input and multiple-output (MIMO) linear system. The UAV dynamics can be decoupled into two modes for the low-level control:

(1) *Longitudinal Mode*: pitch loop

(2) *Lateral Mode*: roll loop.

After dividing the 3D rigid body motion control problem into several loops, cascaded controllers can be designed to accomplish the UAV flight control task. The roll loop control problem or lateral dynamics is carefully studied in this chapter. The roll loop of a UAV can be treated as a SISO (roll-aileron) system around the equilibrium point. In other words, it can be treated as a SISO system around the point where it can achieve a steady-state flight. Steady-state flight means all the force and moment components in the body coordinate frame are constant or zero [13]. An intuitive controller design is to use the classical PID controller structure as follows:

$$C(s) = K_p(1 + \frac{K_i}{s} + K_d s). \tag{4.4}$$

All the controller parameters (K_p, K_i, K_d) will be determined by either off-line or on-line controller tuning experiments.

4.3 ROLL-CHANNEL SYSTEM IDENTIFICATION AND CONTROL

The most intuitive method for roll-channel system identification is to go through an open-loop analysis. However, this method can only be employed with several constraints including small reference (as little as 0.02 rad. for the roll set point [1]) and difficulties in keeping the UAV stable under the open-loop configuration. Therefore, the closed-loop system identification method is used in this chapter because it can guarantee the flight stability of the UAV. The only prior condition is that a rough PID parameter tuning must be performed before the system identification experiment.

The whole system identification procedure includes UAV trim tuning, rough PID tuning to determine $C_0(s)$ and UAV system identification experiments with prespecified excitations, as shown in Fig. 4.1.

Once the system model is derived, another outer-loop controller $C(s)$ will be designed based on modified Ziegler–Nichols (MZN) tuning algorithm or fractional order PI^λ design method, shown in Fig. 4.2.

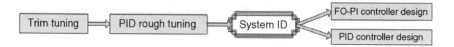

Figure 4.1 FOPI flight controller design procedure.

4.3.1 System Model

For the system model, a simple ARX model is used since the first-order ARX model can provide some simplicity for the further fractional order controller design. The ARX model is defined as

$$\frac{Y(z)}{R(z)} = \frac{a_0 + a_1 z^{-1} + \cdots + a_m z^{-m}}{b_0 + b_1 z^{-1} + \cdots + b_n z^{-n}}, \tag{4.5}$$

where $Y(z)$ is the system output (e.g., the roll angle), and $R(z)$ is the reference signal (e.g., the reference roll angle).

To make a comparison, the first-order plus time delay (FOPTD) model is also simplified via frequency-domain fitting [14] from the high-order ARX model for applying the MZN PID tuning rule,

$$P(s) = \frac{Y(s)}{R(s)} = \frac{Ke^{-Ls}}{Ts + 1}. \tag{4.6}$$

4.3.2 Excitation Signal for System Identification

The excitations for the system ID could be step response, square wave response or PRBS or other prespecified references. The excitation of the system needs also to be carefully chosen because the frequency range of the input reference signal may have a huge impact on the final system identification results. Two reference signals are chosen: square-wave reference and PRBS. PRBS is chosen in this chapter for simulation study because its signal is rich in all the interested frequency.

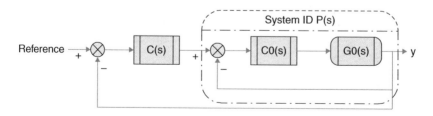

Figure 4.2 System identification procedure.

4.3.3 Parameter Optimization

Least squares error method is used for fitting the model to the real data. Assume the ARX model is given by (4.5). Then

$$
\begin{aligned}
\hat{y}(k) = \frac{1}{b_0}(a_0 r(k) + \cdots + a_m r(k-m)) \\
-b_1 y(k-1) - \cdots - b_n y(k-n)) + e(k),
\end{aligned}
\tag{4.7}
$$

where $e(k)$ is the white noise caused by sensor measurements.

The evaluation function defined below is used to minimize the least squares of the errors

$$
V = \sum_{k=1}^{N} e^T(k)e(k),
\tag{4.8}
$$

where N is the total data length. The classical least squares method can be used here to get the optimal ARX model parameter. In MATLAB, the related function is called `arx` [15]. FOPTD model is simplified from the higher order ARX model using the `getfoptd` function [14].

4.4 FRACTIONAL ORDER CONTROLLER DESIGN

Based on the identified simple model, a new fractional order PI controller is then designed with prespecified performance requirements.

4.4.1 Fractional Order Operators

There are several definitions for fractional-order operators including Riemann–Liouville (RL) definition, Caputo definition, and Grünwald–Letnikov definition. RL definition is one of the most used definitions. The RL fractional integral of function $f(t)$ is defined as [16]

$$
{}_0 D_t^{-\lambda} f(t) \triangleq \frac{1}{\Gamma(\lambda)} \int_0^t (t-\tau)^{\lambda-1} f(\tau) d\tau,
\tag{4.9}
$$

where $0 < \lambda < 1$, $\Gamma(\cdot)$ is the Gamma function defined as

$$
\Gamma(z) = \int_0^\infty e^{-t} t^{z-1} dt, \; Re(z) > 0.
\tag{4.10}
$$

The Laplace transform of the RL fractional integral under zero initial conditions can be derived as

$$\mathscr{L}[_0D_t^{-\lambda}f(t)] = \frac{1}{s^\lambda}F(s), \tag{4.11}$$

where $F(s)$ is the Laplace transform of $f(t)$.

The Caputo fractional integral of order $0 < \lambda < 1$ is defined as [16]

$$_0D_t^{-\lambda}f(t) = \frac{1}{\Gamma(\lambda)}\int_0^t \frac{y(\tau)}{(t-\tau)^{1-\lambda}}d\tau. \tag{4.12}$$

The RL definition and Caputo definition are almost the same except for some initial value settings.

4.4.2 PI^λ Controller Design

With the introduction of fractional derivatives and integrals, the most commonly used PID controller can be extended to $PI^\lambda D^\mu$ controllers with more tuning knobs. PI^λ controller is studied in this chapter since it has the same amount of tuning parameters as the integer order PID controllers to allow a fair comparison. The fractional order proportional integral controller to be designed has the following form of transfer function,

$$C(s) = K_p\left(1 + \frac{K_i}{s^\lambda}\right), \tag{4.13}$$

where $\lambda \in (0, 2)$.

4.4.2.1 Controller Design Specifications

Assume that the open-loop transfer function for the system is given by $G(s)$, the gain crossover frequency is given by ω_c, and phase margin is specified by ϕ_m. To ensure the system stability and robustness, three specifications are proposed as follows [17],

(i) phase margin specification,

$$\text{Arg}[G(j\omega_c)] = \text{Arg}[C(j\omega_c)P(j\omega_c)] = -\pi + \phi_m;$$

(ii) gain crossover frequency specification,

$$|G(j\omega_c)|_{dB} = |C(j\omega_c)P(j\omega_c)|_{dB} = 0;$$

(iii) robustness to gain variation of the plant demands that the phase derivative with respect to the frequency is zero, which is to say that the phase Bode plot is flat around the gain crossover frequency. It means the system is more robust to gain changes and the overshoots of the response are almost unchanged,

$$\frac{d(\text{Arg}(G(j\omega)))}{d\omega}\Big|_{\omega=\omega_c} = 0.$$

4.4.2.2 FOPI Controller Design for the First-Order Systems

To simplify the presentation, the simple form of $G(s)$ is studied in the later part without loss of generality since any complex system can be simplified to a simple model. The typical first-order control plant discussed in this chapter has the following form of transfer function,

$$P(s) = \frac{K}{Ts + 1}. \tag{4.14}$$

Note that, the plant gain K in (4.14) can be normalized to 1 without loss of generality since the proportional factor in the transfer function (4.14) can be incorporated in the proportional coefficient of the controller.

According to the form of the typical first-order systems considered and the FOPI controller discussed, the FOPI controller can be systematically designed following the three specifications introduced above. The FOPI controller parameters can be obtained using the following steps.

The open-loop transfer function $G(s)$ of the fractional order PI controller for the fractional order system is that,

$$G(s) = C(s)P(s).$$

(1) According to the fractional order PI controller transfer function form (4.13), its frequency response could be plotted as follows:

$$C(j\omega) = K_p \left(1 + K_i \omega^{-\lambda} \cos\left(\lambda \frac{\pi}{2}\right) - jK_i \omega^{-\lambda} \sin\left(\lambda \frac{\pi}{2}\right)\right).$$

The phase and gain are as follows:

$$\text{Arg}[C(j\omega)] = -\tan^{-1} \frac{K_i \omega^{-\lambda} \sin(\lambda \pi/2)}{1 + K_i \omega^{-\lambda} \cos(\lambda \pi/2)},$$
$$|C(j\omega)| = K_p J(\omega),$$

where

$$J(\omega) = \left[(1 + K_i \omega^{-\lambda} \cos(\lambda \pi/2))^2 + (K_i \omega^{-\lambda} \sin(\lambda \pi/2))^2\right]^{\frac{1}{2}}.$$

(2) According to the first-order system transfer function (4.14), its frequency response could be plotted as below:

$$P(j\omega) = \frac{1}{T(j\omega) + 1}.$$

The phase and gain of the plant are as follows:

$$\text{Arg}[P(j\omega)] = -\tan^{-1}(\omega T),$$

$$|P(j\omega)| = \frac{1}{\sqrt{1 + (\omega T)^2}}.$$

(3) The open-loop frequency response $G(j\omega)$ is that,

$$G(j\omega) = C(j\omega)P(j\omega).$$

The phase and gain of the open-loop frequency response are as follows:

$$\text{Arg}[G(j\omega)] = -\tan^{-1} \frac{K_i \omega^{-\lambda} \sin(\lambda\pi/2)}{1 + K_i \omega^{-\lambda} \cos(\lambda\pi/2)} - \tan^{-1}(\omega T),$$

$$|G(j\omega)| = \frac{K_p J(\omega)}{\sqrt{1 + (\omega T)^2}}.$$

(4) According to specification (i), the phase of $G(j\omega)$ can be expressed as,

$$\text{Arg}[G(j\omega_c)] = -\pi + \phi_m. \tag{4.15}$$

From (4.15), the relationship between K_i and λ can be established as follows:

$$K_i = \frac{-\tan(\tan^{-1}(\omega_c T) + \phi_m)}{\omega_c^{-\lambda} \sin(\lambda\pi/2) + M}, \tag{4.16}$$

where $M = \omega_c^{-\lambda} \cos(\lambda\frac{\pi}{2}) \tan(\tan^{-1}(\omega_c T) + \phi_m)$.

(5) According to specification (iii) about the robustness to gain variations of the plant,

$$\frac{d(\text{Arg}(G(j\omega)))}{d\omega}\bigg|_{\omega=\omega_c} = \frac{K_i \lambda \omega_c^{\lambda-1} \sin(\lambda\pi/2)}{\omega_c^{2\lambda} + 2K_i \omega_c^{\lambda} \cos(\lambda\pi/2) + K_i^2}$$

$$- \frac{T}{1 + (T\omega_c)^2} = 0. \tag{4.17}$$

From (4.17), the relationship between K_i and λ is:

$$E\omega_c^{-2\lambda} K_i^2 + E + \left[2E\omega_c^{-\lambda} \cos\left(\lambda\frac{\pi}{2}\right) - \lambda\omega_c^{-\lambda-1} \sin\left(\lambda\frac{\pi}{2}\right)\right] K_i = 0,$$

$$E\omega_c^{-2\lambda} K_i^2 + FK_i + E = 0,$$

where $F = 2E\omega_c^{-\lambda}\cos(\lambda\pi/2) - \lambda\omega_c^{-\lambda-1}\sin(\lambda\pi/2)$, then,

$$K_i = \frac{-F \pm \sqrt{F^2 - 4E^2\omega_c^{-2\lambda}}}{2E\omega_c^{-2\lambda}}, \qquad (4.18)$$

where $E = \frac{T}{1+(T\omega_c)^2}$.

(6) From specification (ii), an equation about K_p is:

$$|G_2(j\omega_c)| = |C_2(j\omega_c)P(j\omega_c)| = \frac{K_pJ(\omega_c)}{\sqrt{1+(\omega_cT)^2}} = 1. \qquad (4.19)$$

The above equations (4.17), (4.18), and (4.19) can be solved to get λ, K_i, and K_p. However, the solution may not exist for the integer order PID controller. In other words, fractional order controllers provide a larger solution candidate set compared with integer order ones. A graphical plotting method is used in this chapter to obtain the solutions.

4.4.2.3   FOPI Controller Design for FOPTD Systems

Similarly, the FOPI controller could be designed for the FOPTD systems. The FOPTD system could be modeled by the following:

$$P(s) = \frac{1}{Ts+1}e^{-Ls}. \qquad (4.20)$$

Following the similar derivation described above, the parameters for the FOPI controller could be calculated from the following equations:

$$K_i = \frac{-\tan[\arctan(\omega_cT)+\phi_m+L\omega_c]}{W}, \qquad (4.21)$$

$$A\omega_c^{-2\lambda}K_i^2 + BK_i + A = 0, \qquad (4.22)$$

$$K_p = \frac{\sqrt{1+(\omega_cT)^2}}{J(\omega_c)}, \qquad (4.23)$$

where

$$W = \omega_c^{-\lambda}\cos(\lambda\pi/2)\tan[\arctan(\omega_cT)+\phi_m+L\omega_c]+\omega_c^{-\lambda}\sin(\lambda\pi/2),$$

$$A = \frac{T}{1+(\omega_cT)^2}+L,$$

$$B = 2A\omega_c^{-\lambda}\cos\left(\lambda\frac{\pi}{2}\right)-\lambda\omega_c^{-\lambda-1}\sin\left(\lambda\frac{\pi}{2}\right).$$

4.4.3   Fractional Order Controller Implementation

To implement a PI^λ fractional order controller, an approximation must be used since the fractional order operator has infinite dimensions. The Oustaloup approximation method is to use a band-pass filter to approximate the fractional order controller based on frequency domain response. There are also other methods that directly approximate the FO controller responses [18].

4.4.3.1 Oustaloup Approximation

The Oustaloup Recursive Approximation Algorithm [19] is used in this chapter for simulation part due to its easiness to adapt to MATLAB Simulink environment. Assuming the frequency range is chosen as (ω_b, ω_h), the Oustaloup approximate transfer function for s^γ can be derived as follows:

$$G_{appr}(s) = V \prod_{k=-N}^{N} \frac{s + \omega_k'}{s + \omega_k},$$

where N is a prespecified integer, and the zeros, poles, and the gain can be evaluated from

$$\omega_k' = \omega_b \left(\frac{\omega_h}{\omega_b} \right)^{\frac{k+N+\frac{1}{2}(1-\gamma)}{2N+1}},$$

$$\omega_k = \omega_b \left(\frac{\omega_h}{\omega_b} \right)^{\frac{k+N+\frac{1}{2}(1+\gamma)}{2N+1}},$$

$$V = \left(\frac{\omega_h}{\omega_b} \right)^{-\frac{\gamma}{2}} \prod_{k=-N}^{N} \frac{\omega_k}{\omega_k'}.$$

4.4.3.2 Digital Approximation

However, the Oustaloup approximation can not be directly used in digital control because of digital accuracy issues. s^λ can also be realized by the Impulse Response Invariant Discretization (IRID) method [20] in time domain, where a discrete-time finite dimensional (z) transfer function is computed to approximate the continuous irrational transfer function s^λ, s is the Laplace transform variable, and λ is a real number in the range of $(-1, 1)$. s^λ is called a fractional order differentiator if $0 < \lambda < 1$ and a fractional order integrator if $-1 < \lambda < 0$. This approximation keeps the impulse response invariant.

4.5   SIMULATION RESULTS

The proposed system identification algorithm and fractional-order controller design techniques are first tested on the Aerosim simulation platform, a complete six degrees of freedom UAV dynamic model. For comparison, an integer-order PID controller is also designed using MZN tuning method. Both controllers are tested in scenarios including step response, wind gust response, and payload variation cases. Simulation results verify the advantage of FOC controllers over traditional PID controllers.

4.5.1   Introduction to Aerosim Simulation Platform

Aerosim is a nonlinear six degrees of freedom MATLAB Simulink model designed for the aerosonde UAV [21]. It is developed by Marius Niculescu from u-dynamics with the educational version for free with all the key blocks implemented through dynamic link libraries (dlls).

The control inputs of the aerosonde model include flap, aileron, elevator, rudder, throttle, and the wind. The outputs comprise the following:

- *System States Including Ground Speed:* v_n, v_e, v_d; angular rate: p, q, r; quaternion: q_0, q_1, q_2, q_3; position: p_n, p_e, h, etc.;
- *Sensors Measurements Including GPS:* p_n, p_e, h, v_n, v_e, v_d; inertial measurement unit (IMU): a_x, a_y, a_z, p, q, r; wind: v_n^w, v_e^w, v_d^w; magnetic: h_x, h_y, h_z.

The minimal simulation time step is $0.02s$ (50 Hz).

4.5.2   Roll-Channel System Identification

According to the controller design procedure shown in Fig. 4.1, the trim tuning experiment is performed first in open-loop to get the control input trims for a steady flight state. The trims are $\delta_a = 0$, $\delta_e = -3$ with throttle set as 0.7 (Aerosim does not provide the units for the above variables). It needs to be pointed out that δ_a may not be zero for real UAV platforms due to the manufacturing accuracy. Then the pitch-elevator loop and aileron-roll loop PID controllers should be added with references as shown in Fig. 4.3. For simplicity, the reference pitch angle is set as 0 all the time. The PID parameters are tuned roughly through step response analysis to achieve a steady flight.

A square wave is chosen as the reference input because no sensor noises are added in the simulation. The Steiglitz–Mcbride iteration method is used to get the ARX model of ϕ_{ref}-ϕ loop. Here, time domain system identification method is chosen because the difficulties in choosing the trustable frequency range when analyzing the flight log. MATLAB function stmcb is used to get the models including first-order ARX model, fifth-order ARX model, and FOPTD model simplified from the fifth-order ARX model.

$$G_1(s) = \frac{13.86}{s + 13.76} = \frac{1.0073}{0.0727s + 1}, \tag{4.24}$$

$$G_2(s) = \frac{N_1(s)}{D_1(s)}, \tag{4.25}$$

$$G_3(s) = \frac{1.0336e^{-0.0491s}}{0.0440s + 1}, \tag{4.26}$$

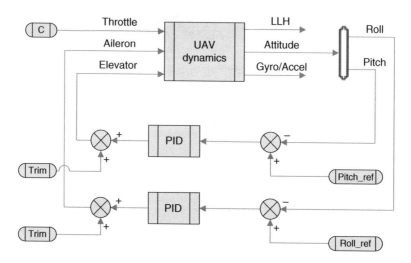

Figure 4.3   UAV flight controller design procedure.

where $N_1(s) = -9.393s^4 + 553.8s^3 + 952.8s^2 + 10960s - 632.9$ and $D_1(s) = s^5 + 21.15s^4 + 662.1s^3 + 1705s^2 + 10920s - 612.3$.

The roll reference $R(N)$ and the roll angle $Y(N)$ are shown together with the simulated square wave responses from the identified models in Fig. 4.4. It can be seen that the simulated time domain responses match the outputs from Aerosim nonlinear model quite accurately for both the first-order and the fifth-order ARX models. The order of five is decided based on numerical experiments.

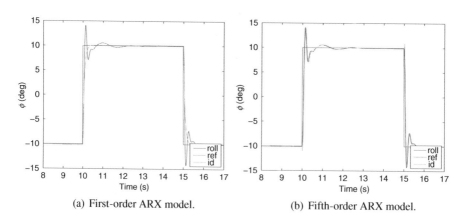

(a) First-order ARX model.   (b) Fifth-order ARX model.

Figure 4.4   Roll-channel system identification.

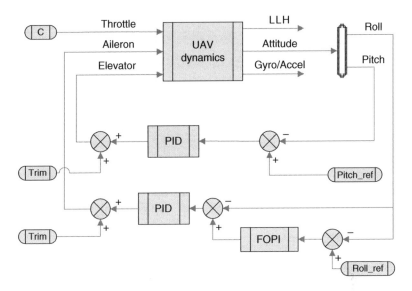

Figure 4.5 FOPI flight controller.

4.5.3 Fractional-Order PI Controller Design Procedure

Given the first-order model identified above, it can be written as $K = 1.0073$ rad^{-1}, $T = 0.0727$ s as (4.14). The fractional-order PI controller to be designed is shown in Fig. 4.5.

The procedure of parameter selection is summarized as below:

(1) The controller performance specifications are chosen as $\omega_c = 10$ rad/s, $\phi_m = 70°$;

(2) The graphic plotting method is used to find the solution for the FOPI parameters. Plot the curve of K_i versus λ according to (4.17), and plot the curve of K_i with respect to λ according to (4.18). The values of λ and K_i can be obtained from the intersection of the two curves, which reads $\lambda = 1.111$, $K_i = 28.31$ rad^{-1};

(3) K_p can be calculated from (4.19), $K_p = 0.5503$ rad^{-1};

(4) Then the designed fractional-order PI controller needs to be validated first, with $K_p = 0.5503$ rad^{-1}, $K_i = 28.31$ rad^{-1}, $\lambda = 1.111$. The Bode plots of the system designed are plotted in Fig. 4.6. It can be seen that the phase Bode plot is flat, at the gain crossover frequency, all three specifications are satisfied precisely.

The Oustaloup realization of FOC controller is used in simulation [19]. The related parameters are chosen as $N = 3$, $\omega_b = 0.05$ rad/s, $\omega_h = 50$ rad/s.

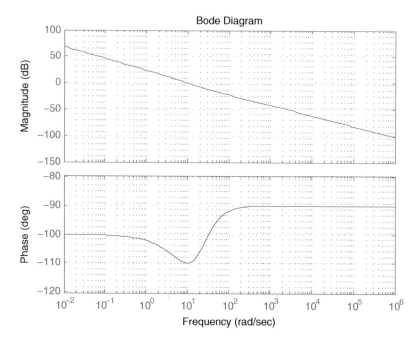

Figure 4.6   Bode plot with designed FOPI controller.

4.5.4   Integer-Order PID Controller Design

As one of the most popular PID controller tuning rules, MZN PID tuning rule is chosen to make a comparison with the designed FOPI controller. MZN tuning method [14] divides the tuning problem into several cases based on different system dynamics:

- Lag Dominated Dynamics ($L < 0.1T$): $K_p = 0.3T/K/L$, $K_i = 1/(8L)$;
- Balanced Dynamics ($0.1T < L < 2T$): $K_p = 0.3T/K/L$, $K_i = 1/(0.8T)$;
- Delay Dominated Dynamics ($L > 2T$): $K_p = 0.15/K$, $K_i = 1/(0.4L)$.

The FOPTD model is identified as $L = 0.0491$ s, $T = 0.0440$ s. It falls into the balanced dynamics category. So, the PID parameters are designed as $K_p = 0.2601$ rad^{-1}, $K_i = 28.4091$ rad^{-1}, $K_d = 0$. The step response comparison ($10°$ for roll tracking) using MZN controller and FOPI controller are shown in Fig. 4.7. It can be observed that the designed FOPI controller responds more quickly and settles faster than the IOPI controller.

4.5.5   Comparison

To show the advantages of FOPI controller over integer-order PID controller, two more experiments were performed to examine the robustness. Wind gusts are very

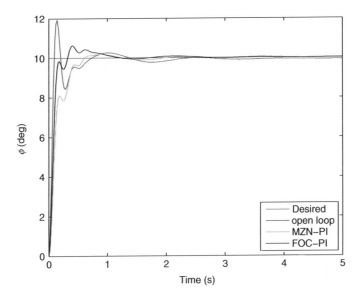

Figure 4.7   Step response comparison: MZN versus FOPI.

common and nontrivial disturbances to the flight control system. Especially for small or micro UAVs, the wind gust can cause crashes if the controller is not well designed. So both FOPI controller and MZN PID controller are tested under extreme conditions when there is a wind gust (10 m/s) lasting for 0.25 s. The results are shown in Fig. 4.8.

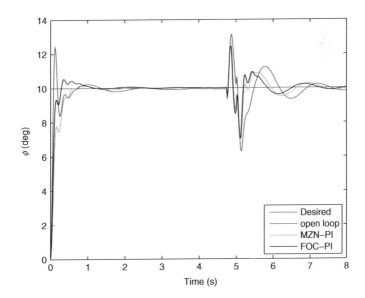

Figure 4.8   Robustness to wind disturbance.

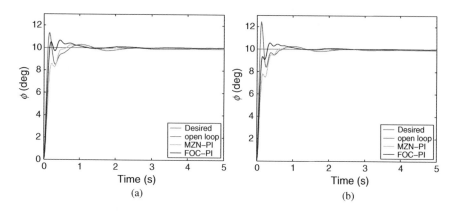

Figure 4.9   Effects of payload gain variations. (a) 80% variation. (b) 120% variation.

It can be seen that the FOPI controller has less overshoot than the MZN PID one and returns to the steady state faster.

Payload variation is also a big issue for small and micro UAVs since the payload can have a big impact on the flight performance. It could be useful if the controller could adapt itself for different sensor payloads. A controller robust to the payload variations could save the UAV end users a lot of time while changing different payloads. To demonstrate the robustness to payload, different controller gains $C1(s)$ are tested with $80\%K$ and $120\%K$ of the original roughly tuned proportional gain, shown in Fig. 4.9. The final step response plots show that the FOPI controller is more robust as compared to the MZN PI controller.

4.6   UAV FLIGHT TESTING RESULTS

Real-flight test results are provided in this section together with the implementation details. The AggieAir2 UAV platform is used to test the proposed fractional order PI controller.

4.6.1   The ChangE UAV Platform

ChangE, an AggieAir UAS platform [22] developed at CSOIS, is used as the experimental platform for the flight controller design and validation [23]. It is built by the authors from the delta-wing RC airframe called Unicorn. The UAV airborne system includes inertial sensors (Microstrain GX2 IMU and u-blox 5 GPS receiver), actuators (elevon and throttle motor), a data modem, an open-source Paparazzi Tiny Twog autopilot, and lithium polymer batteries, as shown in Fig. 4.10. The Microstrain GX2 IMU could provide angle readings (ϕ, θ, ψ) at up to 100 Hz with a typical accuracy

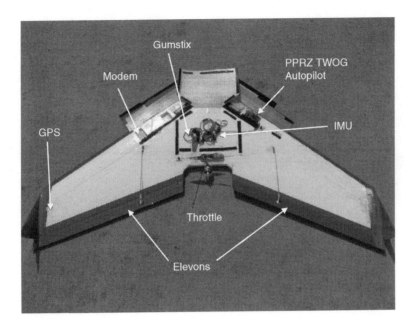

Figure 4.10   ChangE UAV platform.

of ±2° under dynamic conditions [24]. The major specifications of the ChangE UAV are shown in Table 4.1.

The ChangE UAV has both manual RC control mode and autonomous control mode. It communicates with the ground control station (GCS) through a 900-MHz serial modem. The navigation waypoints and flying modes could be changed in real-time from the GCS, shown in Fig. 4.11. The safety pilot could also switch between the manual and autonomous control mode through the RC transmitter in case of emergency. In addition, the Paparazzi GCS software provides on-line parameter changing and plotting functions, which could be easily modified for in-flight tuning of the user-defined controller parameters.

Table 4.1   Change UAS Specifications

ChangE UAV	Specifications
Weight	About 5.5 lb
Wingspan	60 in.
Control inputs	Elevon and throttle
Flight time	≤ 1 h
Cruise speed	15 m/s
Take-off	Bungee
Operational range	Up to 5 miles

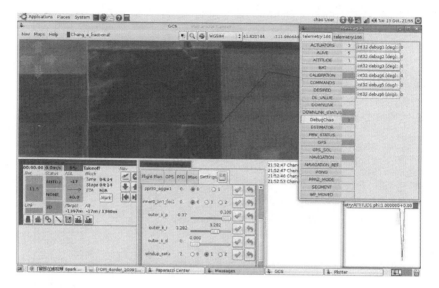

Figure 4.11   Paparazzi GCS.

4.6.2   System Identification

The steady flight tuning is the first step to do a roll-loop system identification. The UAV needs to be manually tuned first to achieve a steady-state flight with zero trim on the elevon at the nominal throttle, which is chosen as 70% based upon the RC flight experiences. The Paparazzi flight controller is replaced by the user designed flight controller (Aggie controller inner loop) both at 60 Hz, as shown in Fig. 4.3. Both the inner roll and pitch PID controllers only include the proportional part. The inner K_p for roll loop is selected as 10038 count/rad, or the maximum value before oscillations is observed by the RC safety pilot. The aileron control inputs are limited within $[-9600, 9600]$ counts. The square response ($[-20°, 20°]$) is generated for the system identification. The reference pitch angle is set as zero all the time. The system response (roll) and the reference roll angle are shown in Fig. 4.12. The square response ($[-10°, 10°]$) is also generated for the model validation, shown in Fig. 4.13.

The first-order ARX model of $\phi_{ref} - \phi$ loop is then calculated based on the flight data log (20 Hz) using least squares algorithm as follows:

$$G(s) = \frac{1.265}{0.901s + 1}.$$

The fifth-order ARX model of $\phi_{ref} - \phi$ loop is also calculated based on the flight log (20 Hz) using least squares algorithm as:

$$G(s) = \frac{N_2(s)}{D_2(s)},$$

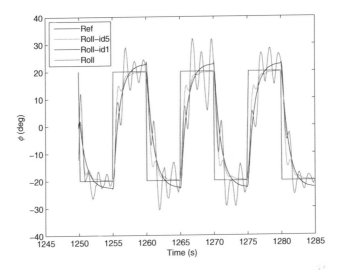

Figure 4.12 Roll-channel system identification ($[-20°, 20°]$).

where $N_2(s) = 0.06108s^5 - 6.825s^4 + 593.2s^3 - 15720s^2 + 220300s + 1071000$
and $D_2(s) = s^5 + 361.5s^4 + 28940s^3 + 136900s^2 + 929000s + 1081000$.

The square wave responses based on the identified model are simulated and plotted together with the real system response in Fig. 4.12. "id5" means the identified fifth-order ARX model and "id1" means the first-order one. It can be seen that the response of the identified model can track the reference signal, and the fifth-order

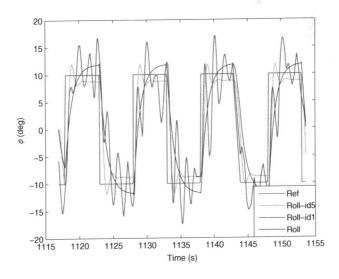

Figure 4.13 Roll-channel system identification ($[-10°, 10°]$).

ARX model identified has better transient responses compared to the first-order ARX model.

The FOPTD model could be calculated from the fifth-order ARX model above using `getfoptd` function [14]:

$$G(s) = 0.9912 \frac{e^{-0.2793s}}{0.3414s + 1}. \tag{4.27}$$

4.6.3 Proportional Controller and Integer Order PI Controller Design

Given the FOPTD model identified above (4.27), a proportional controller could be designed using Ziegler–Nichols tuning rule [14],

$$K_p = \frac{1}{KL/T} = 1.2332.$$

The actual roll tracking result for square reference is shown in Fig. 4.14. It is obvious that the proportional controller has a hard time tracking the roll reference smoothly without overshoots. At the same time, the steady-state tracking error with the designed proportional controller is clearly shown.

Similarly, an integer order PI controller could be designed using MZN tuning rule based upon the identified FOPTD model (4.27),

$$K_p = \frac{0.3T}{KL} = 0.37, \qquad K_i = 0.8T = 3.66.$$

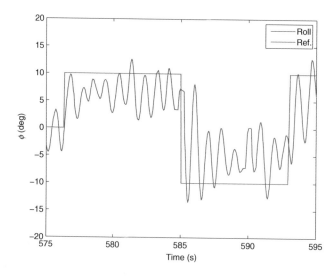

Figure 4.14   Proportional controller for outer roll loop.

The actual step tracking result with this designed integer order PI controller are shown in the later section.

4.6.4 Fractional Order PI Controller Design

The 60-in UAV model is identified as the FOPTD system in (4.27). According to this model, the design procedure of the fractional order PI controller is summarized below,

(1) Given $T = 0.3414$ s, $\omega_c = 1.3$ rad/s, $\phi_m = 80^\circ$;

(2) Plot curve 1, K_i with respect to λ and plot curve 2, K_i with respect to λ based on part I. Obtain the values of λ and K_i from the intersection point on the above two curves, which reads $\lambda = 1.1546$, $K_i = 1.482$;

(3) Calculate the K_p from (4.23), $K_p = 0.8461$;

(4) Then the designed fractional order PI controller can be obtained.

The fractional order part $1/s^{0.1546}$ could be approximated by a fourth-order discrete controller using IRID algorithm (sampling period $T_s = 0.0167$ s) [20]:

$$G(z) = \frac{N(z)}{D(z)},$$

where $N(z) = 0.5203z^4 - 1.1750z^3 + 0.8691z^2 - 0.2245z + 0.0117$, $D(z) = z^4 - 2.4276z^3 + 1.9873z^2 - 0.6062z + 0.0478$.

The Bode plot of $G(z)$ is shown in Fig. 4.15. It can be observed that the fourth-order discrete controller could approximate the frequency response of $1/s^{0.1546}$ around the gain crossover frequency 1.3 rad/s of the open-loop system designed.

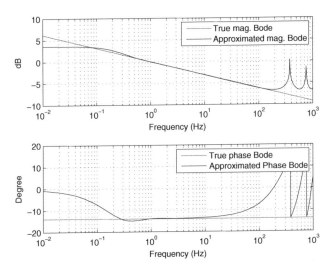

Figure 4.15 Bode plot of $G(z)$.

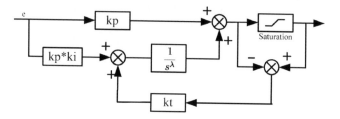

Figure 4.16   Anti-windup for FOPI and IOPI controllers.

An anti-windup block is also added for both the FOPI and IOPI controllers shown in Fig. 4.16. k_t is chosen as $2K_i$.

4.6.5 Flight Test Results

To make a fair comparison between controllers designed using the MZN tuning rule and using the flat phase FOPI tuning rule, the flight tests were conducted for 3 h on October 21, 2009, in the Cache Junction research farm owned by Utah State University. The wind on the ground was predicted to be 0.45–0.9 m/s (1–2 mile/h). Figure 4.17 shows one of the five flight tests for both IOPI and FOPI controllers. The results are fairly repeatable and reproducible. The designed FOPI controller could track the step $10°$ within the sensor resolution range $\pm 2°$ [24]. From Fig. 4.17, it can be observed and concluded that the designed FOPI controller outperforms the designed IOPI controller in both the rise time and overshoot.

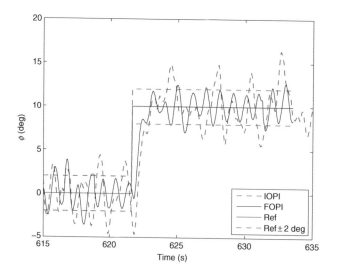

Figure 4.17   FOPI controller for outer roll loop.

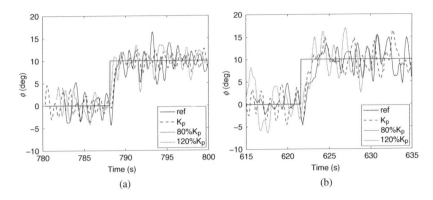

Figure 4.18 FOPI versus IOPI controller with various system gains. (a) FOPI controller. (b) IOPI controller.

The FOPI flight controller is also tested under various system gains to show the robustness of the FOPI controller, shown in Fig. 4.18. It can be observed that the rise time with the FOPI controller is shorter than that with the IOPI controller.

4.7 CHAPTER SUMMARY

In this chapter, the fractional-order proportional integral (FOPI) controller is designed and implemented on the roll loop of a small UAV. To the authors' best knowledge, it is the first fractional-order flight controller that has been implemented to guide the UAV in autonomous flights. Both simulation results and real-flight test data show the effectiveness of the proposed controller design techniques. Future work is to make more comparisons between IOPI and FOPI controller performance for different scenarios such as various wind conditions, various payloads, and the like.

REFERENCES

1. J. Nino, F. Mitrachea, P. Cosynb, and R. D. Keyser. Model identification of a micro air vehicle. *Journal of Bionic Engineering*, 4(4):227–236, 2007.

2. H. Wu, D. Sun, K. Peng, and Z. Zhou. Modeling identification of a micro air vehicle in loitering flight based on attitude performance evaluation. *IEEE Transactions on Robotics and Automation*, 20(4):702–712, 2004.

3. Y. Lee, S. Kim, and J. Suk. System identification of an unmanned aerial vehicle from automated flight tests. In *Proceedings of the AIAA's 1st Technical Conference and Workshop on Unmanned Aerospace Vehicles*, May 2002.

4. G. Cai, B. M. Chen, K. Peng, M. Dong, and T. H. Lee. Modeling and control of the yaw channel of a UAV helicopter. *IEEE Transactions on Industrial Electronics*, 55(9):3426–3434, 2008.

5. S. Duranti and G. Conte. In-flight identification of the augmented flight dynamics of the RMAX unmanned helicopter. In *Proceedings of the Seventeenth IFAC Symposium on Automatic Control in Aerospace*, June 2007.

6. I. Podlubny. Fractional-order systems and $PI^\lambda D^\mu$ controller. *IEEE Transactions on Automatic Control*, 44(1):208–214, 1999.

7. D. Xue, C. Zhao, and Y. Q. Chen. Fractional order PID control of a DC-motor with elastic shaft: a case study. In *Proceedings of the American Control Conference*, pages 3182–3187, June 2006.

8. C. A. Monje, E. Liceaga-Castro, and J. Liceaga-Castro. Fractional order control of an unmanned aerial vehicle (UAV). In *Proceedings of the International Federation of Automatic Control (IFAC) World Congress*, pages 15285–15290, July 2008.

9. C. A. Monje, B. M. Vinagre, V. Feliu, and Y. Q. Chen. Tuning and auto-tuning of fractional order controllers for industry applications. *Control Engineering Practice*, 16(7):798–812, 2008.

10. Y. Q. Chen, T. Bhaskaran, and D. Xue. Practical tuning rule development for fractional order proportional and integral controllers. *Journal of Computational and Nonlinear Dynamics*, 3(2):021403–1–021403–8, 2008.

11. L. Desborough and R. Miller. Increasing customer value of industrial control performance monitoring-Honeywell's experience. In *Proceedings of the 6th International Conference Chemical Process Control*, pages 172–192, July 2001.

12. H. Chao, Y. Cao, and Y. Q. Chen. Autopilots for small unmanned aerial vehicles: a survey. *International Journal of Control, Automation, and Systems*, 8(1):36–44, 2010.

13. B. L. Stevens and F. L. Lewis. *Aircraft Control and Simulation*. 2nd ed. John Wiley & Sons, Hoboken, NJ, 2003.

14. D. Xue and Y. Q. Chen. *Linear Feedback Control: Analysis and Design with MATLAB*. SIAM Press, Philadelphia, PA, 2007.

15. L. Ljung. *MATLAB System Identification Toolbox 7 Users Guide*. The MathWorks, Natick, MA, 2009.

16. I. Podlubny. *Fractional Differential Equations*. Academic Press, San Diego, CA, 1999.

17. H. Li, Y. Luo, and Y. Q. Chen. A fractional order proportional and derivative (FOPD) motion controller: Tuning rule and experiments. *IEEE Transactions on Control System Technology*, 18(2):516–520, 2010.

18. Y. Q. Chen. Applied fractional calculus in controls. In *Proceedings of the American Control Conference*, pages 34–35, June 2009.

19. A. Oustaloup, J. Sabatier, and P. Lanusse. From fractional robustness to CRONE control. *Fractional Calculus and Applied Analysis*, 2(1):1–30, 1999.

20. Y. Q. Chen. Impulse response invariant discretization of fractional order low-pass filters, 2008. http://www.mathworks.com/matlabcentral/fileexchange/21365-impulse-response-invariant-discretization-of-fractional-order-low-pass-filters.

21. M. Niculescu. Aerosim Blockset Users Guide, 2002. http://www.u-dynamics.com.

22. H. Chao, A. M. Jensen, Y. Han, Y. Q. Chen, and M. McKee. In: AggieAir: towards low-cost cooperative multispectral remote sensing using small unmanned aircraft systems. *Advances in Geoscience and Remote Sensing*, IN-TECH, Vukovar, Croatia, 2009.

23. CSOIS. OSAM UAV Web site, 2008. http://www.engr.usu.edu/wiki/index.php/OSAM.

24. Microstrain Inc. Gx2 IMU specifications, 2008. http://www.mirostrain.com.

Chapter 5

Remote Sensing Using Single Unmanned Aerial Vehicle

For the mission of remote sensing of an environmental or agricultural system, the ground-based sensing device can provide a highly accurate mapping but with a limited range (inch-level spatial resolution and second-level temporal resolution). Satellite photos can provide global-level resolution with a large range (30–250 m or lower spatial resolution and week-level temporal resolution). But these photos are expensive and cannot be updated at arbitrary spatial or temporal scales. Small UAVs can provide a high-resolution (meter or centimeter spatial resolution and hour-level temporal resolution) even with an inexpensive camera since most UAVs do not fly so high as satellites.

UAVs equipped with imagers have been used in several agricultural remote sensing applications. High-resolution red–green–blue (RGB) aerial photos can be used to determine the best harvest time of wine grapes [1]. Multispectral images are also shown to be potentially useful for monitoring the ripeness of coffee [2]. However, most current UAV remote sensing applications use large and expensive UAVs with heavy cameras and collect only one band of aerial imagery. As mentioned in Chapter 2, the AggieAir UAS is a low-cost multispectral system for this remote sensing mission. This chapter considers the problem of using a single UAV for the remote sensing missions. First, the remote sensing requirements for water management and irrigation control are discussed, and the problem is divided into two subproblems. The path planning algorithm is introduced to solve the coverage control subproblem. Then, the georeference subproblem is considered with different image stitching and registration techniques. Finally, several typical example missions together with real UAV flight test results are provided including land survey, water area survey, riparian applications, and remote data collection.

Remote Sensing and Actuation Using Unmanned Vehicles, First Edition. Haiyang Chao and YangQuan Chen.
© 2012 by The Institute of Electrical and Electronics Engineers, Inc.
Published 2012 by John Wiley & Sons, Inc.

5.1 MOTIVATIONS FOR REMOTE SENSING

5.1.1 Water Management and Irrigation Control Requirements

The goal of irrigation control is to minimize the water consumption while sustaining the agriculture production and human needs [3]. This optimization problem requires remote sensing to provide real-time feedback from the field including

- *Water*: Water quantities and qualities with temporal and spatial information, for example, the water level of a canal or lake.
- *Soil*: Soil moisture and type with temporal and spatial information.
- *Vegetation*: Vegetation index, quantity and quality with temporal and spatial information, for example, the stage of growth of the crop.

The "real time" here means daily or weekly temporal resolution based on different applications.

5.1.2 Introduction of Remote Sensing

The purpose of remote sensing is to acquire information about the Earth's surface without coming into contact with it. One objective of remote sensing is to characterize the electromagnetic radiation emitted by objects [4]. Typical divisions of the electromagnetic spectrum include the visible light band (380–720 nm), near-infrared (NIR) band (0.72–1.30 μm), and mid-infrared (MIR) band (1.30–3.00 μm). Band-reconfigurable imagers can generate several images from different bands ranging from visible spectra to infrared or thermal bands for various applications. The advantage of an ability to examine different bands is that different combinations of spectral bands can have different purposes. For example, the combination of red–infrared can be used to detect vegetation and camouflage and the combination of red slope can be used to estimate the percent of vegetation cover [2]. After the acquisition of images, further analysis must be made. One widely used processing technique is the normalized difference vegetarian index (NDVI) [5]:

$$\text{NDVI} = \frac{CH_{NIR} - CH_{RED}}{CH_{NIR} + CH_{RED}}, \tag{5.1}$$

where CH_{NIR} and CH_{RED} are spectral reflectance measurements of red and NIR bands. There are also other vegetation index methods, such as the enhanced vegetarian index (EVI), soil-adjusted vegetation index (SAVI) [5], and so on.

The remote sensing problem is defined in detail in Chapter 2. Here only a brief statement is provided. Given an arbitrary area $\Omega \subset R^2$, the goal of remote sensing is to make a mapping from Ω to $\eta_1, \eta_2, \eta_3, \ldots$, representing the density functions for different spectral bands, with preset spatial and temporal resolutions for any position $q \in \Omega$ and any time $t \in [t_1, t_2]$.

5.2   REMOTE SENSING USING SMALL UAVs

Small UAVs are UAVs that can be operated by only one or two people with a flight height less than 10,000 feet above the ground surface. Many of them can be hand-carried and hand-launched. Small UAVs with cameras can easily achieve a meter-level spatial resolution because of their low flight heights. However, this also leads to a smaller footprint size, which represents a limitation on the area that each image can cover. In other words, more georeferencing work is needed to stitch or put images taken at different places together to cover a large-scale water system. Thus, the UAV remote sensing mission can be divided into two subproblems:

- *Coverage Control Problem*: The path planning of the UAV to take aerial images of the whole domain Ω.
- *Georeference Problem*: The registry of each pixel from the aerial images with both temporal and spatial information. For example, the GPS coordinates and time the picture was taken.

There can be two types of solutions. The open-loop solution is to solve the two subproblems separately with the path planning before the UAV flight, and with the image georeferencing after the flight. This method is simple and easy to understand but requires significant experiences to set up all the parameters. The closed-loop solution is to do the path planning and georeference in real-time so that the information from the georeference part can be used as the observer for the path planning controller. This chapter focuses on an open-loop solution.

5.2.1   Coverage Control

Given an arbitrary area Ω, UAVs with functions of altitude and speed keeping and waypoint navigation: speed $v \in [v_1, v_2]$, possible flight height $h \in [h_1, h_2]$; camera with specification: focal length F, image sensor pixel size $PS_h \times PS_v$, image sensor pixel pitch $PP_h \times PP_v$; the interval between images acquired by the camera (the "camera shooting interval") t_{shoot}, the minimal shooting time $t_{shoot_{min}}$, the desired aerial image resolution res; the control objective is

$$\min t_{flight} = g(\Omega, h, v, \{q_1, \ldots, q_i\}, t_{shoot}, res), \tag{5.2}$$

subject to $v \in [v_1, v_2]$, $h \in [h_1, h_2]$, $t_{shoot} = k \times t_{shoot_{min}}$, where t_{flight} is the flight time of the UAV for effective coverage, $g(\Omega, h, v, t_{shoot})$ is the function to determine the flight path and flight time for effective coverage, k is a positive integer.

The control inputs of the coverage controller include bounded velocity v, bounded flight height H, a set of preset UAV waypoints $\{q_1, q_2, \ldots, q_i\}$ and the camera shooting interval t_{shoot}. The system states are the real UAV trajectory $\{\bar{q}_{t_1}, \ldots, \bar{q}_{t_2}\}$ and the system output is a series of aerial images or a video stream taken between t_1 and t_2.

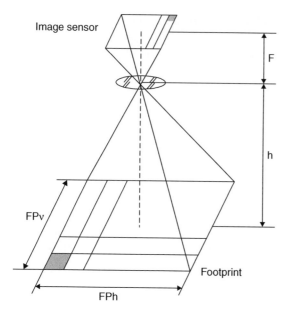

Figure 5.1 Footprint calculation.

Assume that the imager is mounted with its lens vertically pointing down into the Earth; its footprint (shown in Fig. 5.1) can be calculated as

$$FP_h = \frac{h \times PP_h \times PS_h}{F}, \quad FP_v = \frac{h \times PP_v \times PS_v}{F}.$$

Most UAVs can maintain the desired altitude while taking pictures so the UAV flight height h can be determined first based on camera and resolution requirements. Assuming that different flight altitudes have no effect on the flight speed for small UAVs, we get

$$h = \frac{\sqrt{res} \times F}{max(PP_h, PP_v)}. \tag{5.3}$$

Given the flight height h and the area of interest Ω, the flight path, cruise speed, and camera shooting interval must also be determined. Without loss of generality, Ω is assumed to be a rectangular since most other polygons can be approximated by several smaller rectangles. The most intuitive flight path for UAV flight can be obtained by dividing the area into strips based on the group spatial resolution, shown in Fig. 5.2a. The images taken during UAV turning are not used for our remote sensing missions because all the aerial images should have similar resolutions for georeferencing simplicities. Due to the limitation from the UAV autopilot, GPS accuracy and wind, the small UAV cannot follow a straight line perfectly along the preset waypoints. To compensate the overlapping percentage between two adjacent sweeps, o must also be determined before flight; this compensation is based on experiences from the later image stitching as shown in Fig. 5.2b [6].

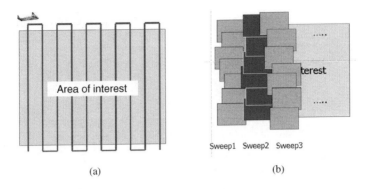

Figure 5.2   UAV flight path [6]. (a) Ideal flight path. (b) Flight path with overlap.

Given the overlapping percentage $o\%$ between sweeps, the ground overlapping o_g can be determined by

$$o_g = (1 - o\%) \times FP_h. \tag{5.4}$$

The minimal camera shooting interval can be computed as

$$t_{shoot_{min}} = \frac{(1 - o\%) \times FP_v}{v}. \tag{5.5}$$

This open-loop solution is intuitive, robust to all the polygons, and requires little computations. However, this method requires that many parameters, especially the overlapping percentage $o\%$ to be set up based on experiences; it cannot provide an optimal solution. More work on a closed-loop real-time solution is needed for an optimal solution.

5.2.2   Georeference Problem

After the aerial images are taken and sent back to the ground, postflight image processing is needed since the UAV cannot maintain perfectly level flights all the time [7,8].

Problem Statement for Post Flight Image Stitching

Given the aerial images stream $\{I_1, I_2, \ldots, I_m\}$ and the UAV flight data logger stream $\{t_1, \ldots, t_n\}$, $\{q_1, \ldots, q_n\}$, $\{\psi_1, \ldots, \psi_n\}$, $\{\theta_1, \ldots, \theta_n\}$, $\{\phi_1, \ldots, \phi_n\}$, map $\eta(R^3, \psi, \theta, \phi, t)$ back to $\eta(R^2, 0, 0, \phi, t)$, where ψ, θ and ϕ represent the roll, pitch, and yaw angles, respectively.

This problem can be solved by feature-based mapping, or GIS-based mapping, or both.

5.2.2.1   Feature-Based Stitching

With feature-based stitching, the aerial images are stitched together based on some common features such as points, lines, or corners. These features can either be

Figure 5.3   Photos before stitching [6]. (*See insert for color representation of this figure.*)

specified manually by humans or recognized automatically by some algorithms. For example, PTGui software needs reference points to perform image stitching [9]. This method works well for photos sharing several common features. A simple example is shown in Figs. 5.3 and 5.4 [6]. The aerial photos were taken at about 150 m above

Figure 5.4   Photos after stitching (courtesy of Marc Baumann). (*See insert for color representation of this figure.*)

the ground by a GF-DC mounted on an RC airplane controlled by a human operator. The problem with this method is that the photos are not georeferenced, and with no absolute coordinates to tie to. The error can keep increasing with the image numbers. Because of the small image footprint of small UAVs and the lack of permanent distinguishing features of most agricultural fields, it is almost impossible to find enough common features in every picture to correctly tie the photos together. There are also feature-based stitching methods for image georeferencing. These take features on each photo and compare them to photos, which already have GIS information. This method performs the stitching task better because each photo is georeferenced and has some absolute coordinates to tie to. However, the photos with the GIS information are often taken once a year. For some photos taken from the UAV, they are not current enough to find similar features.

5.2.2.2 UAV Position and Attitude-Based Stitching

This method uses the data from the UAV including q, θ, ϕ, ψ to map the image back to the related ground frame and registry each pixel with its ground coordinates and electromagnetic density measurements. The advantage is that this method uses all the information from the UAV and can guarantee a bounded global error even for the final big image. However, it requires perfect synchronization between the aerial images and the attitude logs, which means the full authority in UAV autopilot and sensor package. The inability to access the autopilot code is one of the disadvantages of using off-the-shelf UAVs for remote sensing missions.

A three-dimensional (3D) mesh M_{hw} must first be generated in F_{cam} [6], which describes the spatial distribution of its corresponding image. This mesh has $n \times n$ elements, illustrated in Fig. 5.5 [10]. Each element of (M_{hw}) can be calculated as

$$q(w, h)_{cam} = \begin{bmatrix} x(w, h)_{cam} \\ y(w, h)_{cam} \\ f \end{bmatrix}, \forall q(w, h)_{cam} \in M_{wh}, \tag{5.6}$$

$$x(w, h)_{cam} = \frac{S_w(2w - n)}{2n}, \quad y(w, h)_{cam} = \frac{S_h(2h - n)}{2n},$$

where $w, h \in [1, n]$, f is the focal length of the camera, w is the column index of the mesh, h is the row index of the mesh, S_w and S_h are the width and height of the image sensor, respectively.

The mesh must be rotated first with respect to F_{body}, F_{nav} and then translated to the ground frame F_{gps}. The manner in which the camera is mounted on the UAV body decides the rotation matrix R_{body}^{cam}, which can be a constant or a dynamic matrix. The angles (yaw, pitch, and roll ψ_c, θ_c, ϕ_c) describe the orientation of the camera with respect to the body frame.

$$R_{body}^{cam} = (R_{cam}^{body})^T = (R_{zyx}(\psi_c + 90, \theta_c, \phi_c))^T, \quad R_{body}^{cam}, R_{cam}^{body} \in SO(3). \tag{5.7}$$

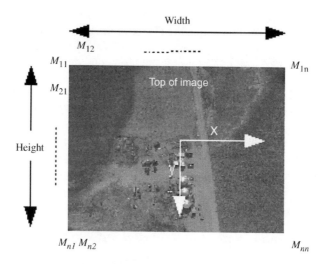

Figure 5.5 Camera frame and mesh [6]. (*See insert for color representation of this figure.*)

The rotation matrix from F_{body} to F_{nav} can be determined from the orientation of the aircraft (ψ, θ, ϕ) with respect to the navigation coordinate system, which can be measured by the UAV onboard sensors.

$$R_{nav}^{body} = \begin{bmatrix} c_\theta c_\psi & -c_\phi s_\psi + s_\phi s_\theta c_\psi & s_\phi s_\psi + c_\phi s_\theta c_\psi \\ c_\theta s_\psi & c_\phi c_\psi + c_\phi s_\theta s_\psi & -s_\phi c_\psi + c_\phi s_\theta s_\psi \\ -s_\theta & s_\phi c_\theta & c_\phi c_\theta \end{bmatrix}, \quad R_{nav}^{body} \in SO(3), \quad (5.8)$$

where c_θ, s_θ stand for $\cos(\theta)$ and $\sin(\theta)$, respectively.

The rotation transformations can be then calculated with all the above rotation matrices as follows:

$$q_{nav} = R_{nav}^{body} \; R_{body}^{cam} \; q_{cam}, \quad (5.9)$$

The mesh now represents the location of the image sensor in the Navigation Frame F_{nav}. To represent the location of the picture on the Earth, the vectors in the mesh are scaled down to the ground. It is assumed that the Earth is flat because each picture has such a small footprint. q_{nav} is the element of the new projected mesh, $q_{nav}(z)$ is the z component of the element in the unprojected mesh, and h is the height of the UAV when the picture was taken.

$$q_{nav}^P = \frac{h}{q_{nav}(z)} q_{nav} \quad (5.10)$$

The elements of the mesh are now rotated into ECEF coordinates using the latitude (λ) and the longitude (φ) of the UAV when the picture was taken. Then the mesh is translated by the position vector ($\vec{P}$) of the UAV in ECEF coordinates.

$$R_{gps}^{nav} = R_{yyz}(-\lambda, 90°, \varphi), \qquad R_{gps}^{nav} \in SO(3), \tag{5.11}$$

$$q_{gps} = R_{body}^{nav} \; q_{nav}^{P} + \vec{P}. \tag{5.12}$$

After the above calculations, the meshes can be further processed by some 3D imaging software such as NASA World Wind [11]. NASA World Wind can place the meshes correctly on the Earth and provide an interactive 3D displaying.

5.3 SAMPLE APPLICATIONS FOR AGGIEAIR UAS

Experimental results are shown in this section to demonstrate the effectiveness of the proposed algorithm for remote sensing missions using AggieAir, a small UAV. Several sample applications are introduced in detail. First, real-time surveillance and farmland coverage missions are introduced. Acquisition of photographic data for road survey, water area coverage, and riparian survey are illustrated involving the use of both the RGB and NIR imagers. Finally, a remote data collection application is offered to demonstrate the feasibility of using UAVs to collect data from ground-based sensors through wireless modems.

5.3.1 Real-Time Surveillance

Real-time surveillance is a very important application of small UAVs especially under emergent situations. Scenarios such as forest fire, chemical leaking search and rescue all require fast surveillance techniques without jeopardizing human lives. AggieAir type UAS is an ideal platform for such tasks since it can be quickly deployed without depending on the runway. Real-time aerial images collected by AggieAir can be transmitted to the ground with all the position and orientation information. Then human operators can identify the target and provide useful georeferenced information with the developed image georeferencing subsystem. An example case is the mission of AUVSI student UAS competition in 2009. The UAS is expected to find out information of several unknown targets within a prespecified area as quick as possible. The interested information includes accurate GPS position, target color, and target orientation. The search area is about 1 square mile big. AggieAir Tiger successfully finished the preplanned flight plan and locates all the targets following the algorithm provided in the former sections. An example aerial image with a target (F) is shown in Fig. 5.6. The image was taken at 91.99 m above the ground with the camera orientation of 15.6 degree roll, 13.0 degree pitch, and -171.4 degree yaw.

5.3.2 Farmland Coverage

This irrigation optimization problem requires remote sensing to provide real-time feedback from the farmland field including water, soil, and vegetation. The "real-time"

Figure 5.6 Example target image acquired by AggieAir.

here means daily or weekly temporal resolution based on different applications. The UAV is needed to measure the soil moisture of the area to help save water. More importantly, the ground probes and Landsat data can be used to calibrate the images from small UAVs using the downscaling techniques [12,13]. After calibration, these images should be able to measure soil moisture and evapotranspiration for water managers and farmers whenever it is needed. A research farm (1 square mile) coverage map is provided in Fig. 5.7 to show the capability of AggieAir.

Figure 5.7 Cache Junction farm coverage map (courtesy of A. M. Jensen).

5.3.3 Road Surveying

AggieAir UAS can also provide low-cost aerial images for road and highway construction and maintenance. Figure 5.8 shows a highway intersection located in Logan Canyon, which was rebuilt in 2008 to improve safety for drivers turning onto the main road. The Utah Department of Transportation (UDOT) provides the picture taken by

(a)

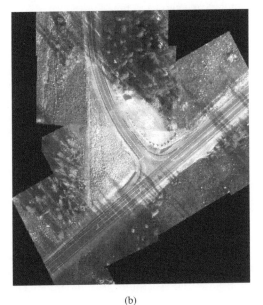

(b)

Figure 5.8 Beaver resort intersection (courtesy of A. M. Jensen). (a) Before construction (manned aircraft picture from UDOT). (b) After construction (AggieAir picture). (*See insert for color representation of this figure.*)

a manned aircraft before construction, shown in the left of Fig. 5.8. The aerial images acquired by AggieAir is shown in the right. It can be seen that the aerial images from AggieAir have a better resolution with feature-based stitching.

5.3.4   Water Area Coverage

Water areas include wetlands, lakes, ponds, and the like. Water areas could provide lots of information to ecological environment changes, flood damage predictions, and water balance management. Desert Lake coverage mission is a typical example, which lies in west-central Utah (latitude: $39°22'5''N$, longitude: $110°46'52''W$). It is formed from return flows from irrigated farms in that area. It is also a waterfowl management area. This proposes a potential problem because the irrigation return flows can cause the lake to have high concentrations of mineral salts, which can affect the waterfowl that utilize the lake. Managers of the Desert Lake resource are interested in the affect of salinity control measures that have been recently constructed by irrigators in the area. This requires estimation of evaporation rates from the Desert Lake area, including differential rates from open water, wetland areas, and dry areas. Estimation of these rates requires data on areas of open water, wetland, and dry lands, which, due to the relatively small size and complicated geometry of the ponds and wetlands of Desert Lake, are not available from satellite images. A UAV can provide a better solution for the problem of acquiring periodic information about areas of open water, etc., since it can be flown more frequently and at little cost.

The whole Desert Lake area is about 2×2 miles. It comprises four ponds and some wetland areas. The early version of AggieAir imaging payload, GF-DV, is used in this mission together with real-time, simultaneous RGB and NIR videos. Both the RGB and NIR videos are transmitted back to the ground station in real time. The photos collected by AggieAir are stitched and shown in Fig. 5.9.

5.3.5   Riparian Surveillance

Riparian buffer surveillance is becoming increasingly more important since it is challenging to maintain stream ecosystem integrity and water quality with the current rapidly changing land use [14]. AggieAir UAS could be used in several applications including river tracking, vegetation mapping, hydraulic modeling, and so on.

5.3.5.1   River Tracking

The path and flow of a river might constantly change due to drought, flood, or other natural calamities. Because of this, the aerial images of the river path could be outdated or in low quality, making it difficult to perform studies of the changed river and the variations of its nearby ecological system. AggieAir UAS platform with the high-resolution multispectral camera system could present a real-time low-cost solution to the river tracking problem [15]. A flight plan with 3D waypoints could be formed by integrating flow line data from NHDPlus (National Hydrography Dataset

(a)

(b)

Figure 5.9 Desert lake coverage map
(courtesy of A.M. Jensen). (a) RGB image.
(b) NIR image. (*See insert for color
representation of this figure.*)

Plus) and DEM (Digital Elevation Model) from USGS (U.S. Geological Survey). The images captured by the cameras are processed in real time. Based on the information derived from these images, waypoints are dynamically generated for the autonomous navigation so that the UAV can exactly follow the changed river path and the focus of each image from the camera system is on the center of the river. The actual flight results collected in several flying experiments along a river verify the effectiveness of AggieAir System, shown in Fig. 5.10. Example RGB and NIR pictures acquired by AggieAir are also shown in Fig. 5.11.

5.3.5.2 Vegetation Mapping and Hydraulic Modeling

In addition to agricultural applications, AggieAir is also a useful platform for riparian projects. Figure 5.12 shows some imagery taken with AggieAir of a small section of

Figure 5.10 River tracking map after stitching (courtesy of A.M. Jensen). (*See insert for color representation of this figure.*)

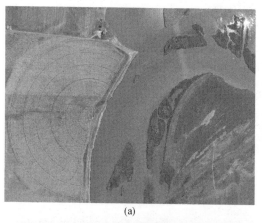

(a)

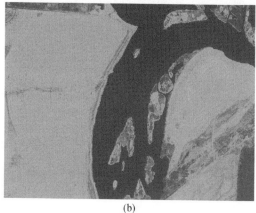

(b)

Figure 5.11 Sample pictures for river tracking. (a) RGB image. (b) NIR image. (*See insert for color representation of this figure.*)

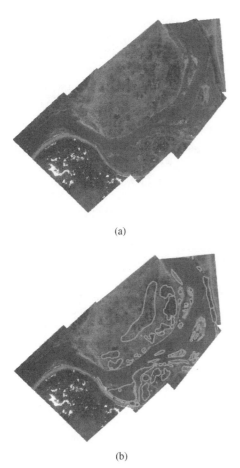

(a)

(b)

Figure 5.12 Oneida Narrows imagery (courtesy of A.M. Jensen and S. Clemens for image processing). (a) Substrate map. (b) Vegetation map. (*See insert for color representation of this figure.*)

the Oneida Narrows near Preston, Idaho. A team of engineers used this imagery to map the substrate and vegetation for 2D hydraulic and habitat modeling. Normally, the team uses low-resolution, outdated imagery to map rivers. This can be difficult when the vegetation, the path and the flow of the river are always changing. The imagery from AggieAir, however, was up-to-date (within a week) and had high spatial resolution (5 cm), which made mapping the river quick and easy. Not only could different types of vegetation be distinguished from the imagery, but different types of sediment, like sand piles, could also be easily distinguished.

5.3.6 Remote Data Collection

Many agricultural and environmental applications require deployment of sensors for measurement of the interested field. However, it is not always easy or inexpensive to collect all the data from remote data loggers for further processing. Many applications still require humans to get close to the ground-based sensors to retrieve the data from their data loggers. Wireless sensor networks and satellite networks are used in environmental data collection applications, but wireless communication can be expensive and vulnerable to changing environmental conditions (such as loss of line-of-sight due to vegetation growth). This problem is especially difficult when the sensors are deployed sparsely over a large geographic area where transportation might be limited by terrain conditions. UAVs can fly into such areas without affecting the vegetation on the ground; they can spare humans from having to enter dangerous or difficult areas; and they might be able to operate at lower costs that might be required for approaches involving direct human access to the data. Moreover, UAVs can achieve better wireless communication since the signal can be transmitted more dependably in the air than near ground level.

One preliminary experiment was performed to test the feasibility to use UAVs for remote data collection. The CR 206 data loggers from Campbell Scientific Inc. are used as both ground sensors and communication modules. The CR206 data logger is a small measurement and control unit with a 12-bit A/D converter, an on-board 915 MHz spread spectrum radio modem, and 512 kbytes memory [16]. It can transmit data to another logger with the communication range of about 300 m on the ground. Four CR 206 data loggers were sparsely spread out on a research farm with more than 300 m between them, shown in Fig. 5.13. One CR206 data logger was mounted at the bottom of the UAV as a remote data collector, which scans every 15 s and sends data out if it detects another CR206. The preplanned path is to fly above all the four data loggers and circle around, as shown in Fig 5.13(a). The UAV will first start from take off point, climb to the desired height and begin following waypoints 1–2–3–4–5. The UAV will repeat this cycle 5–1 until the battery is exhausted. The flight height is set as 100 m and the cruise speed for the UAV is set to 14 m/s. The UAV flight trajectory is plotted in Fig. 5.13a with time steps marked. The UAV has a data logging rate of 3 Hz.

The UAV actually flew two rounds from 10:35:16 AM to 10:40:12 AM. The data receiver got no signal during the take-off period, which is from 10:34 AM to 10:35:16 AM as shown in Fig. 5.13a. Data collected from data logger 3 are analyzed in detail with the signal strength and the transmission distance shown in Fig. 5.14.

It is clear that the UAV can retrieve data given the UAV stays close to the ground station as shown in Fig. 5.14. But there are some obvious delays caused by the UAV flight speed. It is also observed that the signal strength can vary with the distance. The maximal distance to successfully retrieve data is as far as about 800 m, about three times the range on the ground. However, it is also observed that the communication is highly affected by the height and the surroundings of the data logger. Data logger 4 is placed in the grass about 2 feet high, which results in bad communication. Data loggers 1 and 5 have a good communication because of their position above the ground. Further tests with a wider ground sensor distribution, as far as five miles, will be attempted. One more function, that of loitering or circling around the ground sensors, will also be added.

5.3.7   Other Applications

Another typical example of remote data collection is fish tracking. In order to understand fish habitats, radio transmitters are planted in fishes in order to locate and track

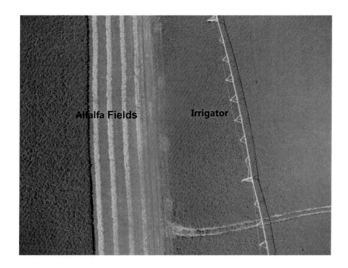

Figure 1.4 Typical agricultural field (Cache Junction, UT).

Figure 1.6 Fog evolution (taken in Yellowstone National Park).

Remote Sensing and Actuation Using Unmanned Vehicles, First Edition. Haiyang Chao and YangQuan Chen.
© 2012 by The Institute of Electrical and Electronics Engineers, Inc.
Published 2012 by John Wiley & Sons, Inc.

Figure 5.3 Photos before stitching [6].

Figure 5.4 Photos after stitching (courtesy of Marc Baumann).

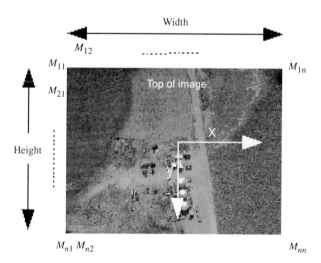

Figure 5.5 Camera frame and mesh [6].

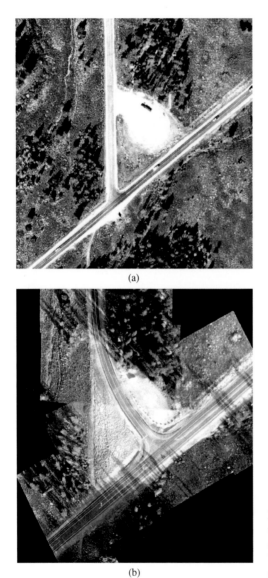

(a)

(b)

Figure 5.8 Beaver resort intersection (courtesy of A.M. Jensen). (a) Before construction (manned aircraft picture from UDOT). (b) After construction (AggieAir picture).

(a)

(b)

Figure 5.9 Desert lake coverage map (courtesy of A.M. Jensen). (a) RGB image. (b) NIR image.

Figure 5.10　River tracking map after stitching (courtesy of A.M. Jensen).

(a)

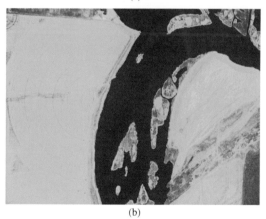

(b)

Figure 5.11 Sample pictures for river tracking. (a) RGB image. (b) NIR image.

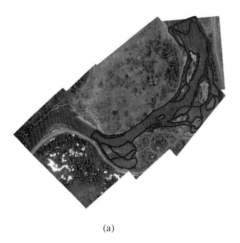

(a)

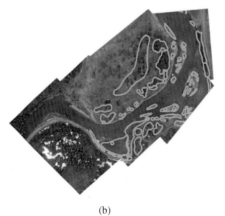

(b)

Figure 5.12 Oneida Narrows imagery (courtesy of A.M. Jensen and S. Clemens for image processing). (a) Substrate map. (b) Vegetation map.

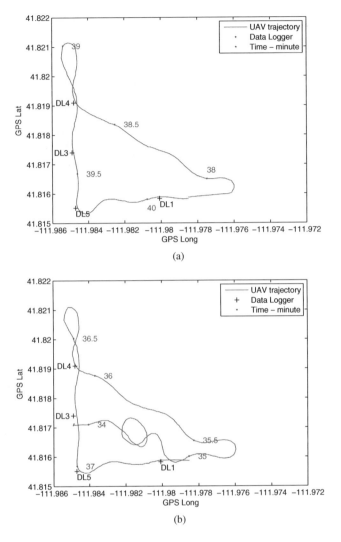

Figure 5.13   UAV trajectory. (a) Trajectory first round. (b) Trajectory second round.

their movements. Human operators are needed to drive a boat around a lake or down a river following the periodic beacon sent from the transmitters. The beacon is heard through a radio receiver with a directional antenna and its strength is highly dependent on the distance and the direction the antenna is pointed at. AggieAir could be employed here with an onboard self-designed device to catch the signal from the transmitter and record its strength. Thus, the location of the fish can be found and recorded much easier and faster since the wireless signal transmits much better in the air.

There are also many other potential applications for small UAV-based remote sensing. One is for thermal image applications such as search and rescue, wildlife surveying. A Photon 320 thermal camera was integrated with AggieAir UAS for

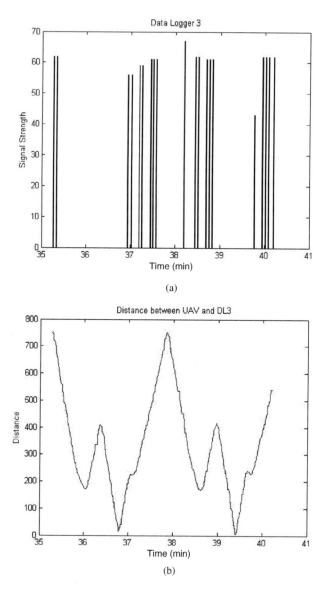

Figure 5.14   Further analysis of data logger 3. (a) Signal Strength from DL3. (b) Distance between UAV and DL3.

prototype validations [17]. The sample thermal image taken on the ground is shown in Fig. 5.15. In fact, the white color on the picture means a high temperature while the black color means a low temperature. It can be observed that the vehicle had higher temperature than the human since the air conditioning was running. Similarly, a sample thermal image taken in the air is shown in Fig. 5.15b. It is worth mentioning

(a)   (b)

Figure 5.15 Sample thermal pictures. (a) Ground thermal picture. (b) Aerial thermal picture.

in Fig. 5.15b that the engine of the back vehicle was shut off at the moment the picture is taken and the front vehicle was left on all the time.

5.4 CHAPTER SUMMARY

This chapter characterizes the problem of using UAVs for remote sensing in water management and irrigation control applications, and provides an outline of a band reconfigurable solution to the problem. The big problem is divided into two subproblems: the path planning subproblem and the image georeferencing subproblem for separate solutions. Several representing experimental results show the effectiveness of the proposed solution.

REFERENCES

1. L. F. Johnson, S. R. Herwitz, S. E. Dunagan, B. M. Lobitz, D. V. Sullivan, and R. E. Slye. Collection of ultra high spatial and spectral resolution image data over California vineyards with a small UAV. In *Proceedings of the International Symposium on Remote Sensing of Environment*, November 2003.
2. L. F. Johnson, S. R. Herwitz, B. M. Lobitz, and S. E. Dunagan. Feasibility of monitoring coffee field ripeness with airborne multispectral imagery. *Applied Engineering in Agriculture*, 20:845–849, 2004.
3. S. Z. Fedro, G. S. Allen, and A. C. Gary. Irrigation system controllers. *Series of the Agricultural and Biological Engineering Department, Florida Cooperative Extension Service, Institute of Food and Agricultural Sciences, University of Florida*, 1993.
4. B. C. James. *Introduction to Remote Sensing*. 4th ed. Guilford Press, New York, 2006.
5. W. R. Terrill. A FAQ on vegetation in remote sensing, 1993. http://www.tetracam.com.
6. H. Chao, M. Baumann, A. M. Jensen, Y. Q. Chen, Y. Cao, W. Ren, and M. McKee. Band-reconfigurable multi-UAV-based cooperative remote sensing for real-time water management and distributed irrigation control. In *Proceedings of the International Federal of Automatic Control (IFAC) World Congress*, pages 11744–11749, July 2008.

7. R. P. Randy and P. Sirisha. Development of software to rapidly analyze aerial images. In *Proceedings of the American Society of Agricultural and Biological Engineers Annual International Meeting*, July 2005.

8. H. Xiang and L. Tian. Autonomous aerial image georeferencing for an UAV based data collection platform using integrated navigation system. In *Proceedings of the American Society of Agricultural and Biological Engineers Annual International Meeting*, June 2007.

9. PTGui Developers. PTGui stitching software, 2007. www.ptgui.com.

10. A. M. Jensen. gRAID: a geospatial real-time aerial image display for a low-cost autonomous multispectral remote sensing platform. Master's thesis, Utah State University, 2009.

11. NASA. World wind software, 2007. http://worldwind.arc.nasa.gov/index.html.

12. Y. H. Kaheil, M. K. Gill, M. McKee, L. A. Bastidas, and E. Rosero. Downscaling and assimilation of surface soil moisture using ground truth measurements. *IEEE Transactions on Geoscience and Remote Sensing*, 46(5):1375–1384, 2008.

13. Y. H. Kaheil, E. Rosero, M. K. Gill, M. McKee, and L. A. Bastidas. Downscaling and forecasting of evapotranspiration using a synthetic model of wavelets and support vector machines. *IEEE Transactions on Geoscience and Remote Sensing*, 46(9):2692–2707, 2008.

14. S. J. Goetz. Remote sensing of riparian buffers: past progress and future prospects. *Journal of the American Water Resources Association*, 42(11):133–143, 2006.

15. Y. Han, H. Dou, and Y. Q. Chen. Mapping river changes using low cost autonomous unmanned aerial vehicles. In *Proceedings of the American Water Resources Association Spring Specialty Conference on Managing Water Resources Development in a Changing Climate*, May 2009.

16. Campbell Scientific. CR200 series brochure, 2005. http://www.campbellsci.com/2/20354/14/1.

17. H. Sheng, H. Chao, C. Coopmans, J. Han, M. McKee, and Y. Q. Chen. Low-cost UAV-based thermal infrared remote sensing: platform, calibration and applications. In *Proceedings of the IEEE/ASME International Conference on Mechatronics and Embedded Systems and Applications*, pages 38–43, July 2010.

Chapter 6

Cooperative Remote Sensing Using Multiple Unmanned Vehicles

The limited power capacity of small UAVs creates challenges while using UAVs for remote sensing missions of a large-scale system. This motivates the group use of small UAVs since the aerial image collection task can be finished in parallel on different small UAVs. For instance, some irrigation applications may require remote sensing of a land as big as 30 square miles within 1 h. Acquisition of imagery on such a geographic scale is difficult for a single small UAV. Instead, groups of small UAVs can solve this problem because they can provide images in a shorter time even with more spectral bands since UAVs with configurable spectral bands can work in groups. Of course, the low costs make it possible to use small UAVs in large amounts for civilian applications.

Following the similar definition of the remote sensing problem in Chapter 5, the following missions will need multiple UAVs being operated cooperatively:

- Measure multispectral images $\eta_1, \eta_2, \eta_3, \ldots$ simultaneously. For example, images from multiple spectral bands can provide far more information and they are indispensable for applications such as soil moisture estimation.

- Measure $\eta_i(q, t)$ within a short time. Some water applications may require images that are taken when the sunshine is the strongest. Small UAVs could work in groups for these kinds of missions to ensure the mission accomplishment within a short time.

- Measure the gradient in real time. Applications such as monitoring of a fast-evolving process may require information like the gradient or flow rate in real time. For instance, the gradient of the temperature and wind field in a forest fire scenario can greatly affect the prediction of the fire spread.

Remote Sensing and Actuation Using Unmanned Vehicles, First Edition. Haiyang Chao and YangQuan Chen.
© 2012 by The Institute of Electrical and Electronics Engineers, Inc.
Published 2012 by John Wiley & Sons, Inc.

The above typical multi-UAV missions can be categorized into two groups: formation-based applications and non-formation-based ones. The consensus-based formation control algorithms are looked into carefully in this chapter with theoretical approaches and experimental validations on MASnet hardware simulation platform. The wind profiling measurement problems is also considered using small UAVs. Similar formation control techniques can be applied on UAVs for wind field measurement.

6.1 CONSENSUS-BASED FORMATION CONTROL

The basic idea of consensus algorithm is to achieve a common state among groups of agents through multivehicle communications [1]. For example, the rendezvous mission requires all the agents to converge to a same position with no preknown agreements. The rendezvous mission is a base for other formation control tasks since the offset can be added to the rendezvous mission to achieve the axial alignment or formation control. The basics of the consensus algorithm and its implementation on our MASnet hardware simulation platform are introduced in this section. The motivation here is to apply this multivehicle consensus algorithm on real hardware platforms for the formation control.

6.1.1 Consensus Algorithms

Assume that the unmanned vehicles have single-integrator dynamics given by

$$\dot{\xi}_i = u_i, \quad i = 1, \ldots, n, \tag{6.1}$$

where $\xi_i \in \mathbb{R}^m$ is the state of the ith vehicle (e.g., position), and $u_i \in \mathbb{R}^m$ is the control input (e.g., velocity). The consensus algorithm is proposed as

$$u_i = -\sum_{j \in \mathcal{J}_i(t)} k_{ij}(t)(\xi_i - \xi_j), \quad i = 1, \ldots, n, \tag{6.2}$$

where $\mathcal{J}_i(t)$ represents the set of vehicles sharing states with vehicle i at time t, and $k_{ij}(t)$ is a positive weighting factor at time t [1–5]. $\mathcal{J}_i(t)$ combined with $k_{ij}(t)$ can also be called a Laplacian matrix $L(t) \in \mathbb{R}^{n \times n}$ in the graph theory. $L(t) = [\ell_{ij}(t)]$ is defined as follows:

$$\ell_{ij}(t) = \begin{cases} \sum_{j \in \mathcal{J}_i(t) \setminus \{i\}} k_{ij}(t), & \text{if } i = j; \\ 0, & \text{if } i \neq j, j \notin \mathcal{J}_i(t); \\ -k_{ij}(t), & \text{if } i \neq j, j \in \mathcal{J}_i(t) \setminus \{i\}. \end{cases} \tag{6.3}$$

For the control law described in (6.2), consensus is said to be reached asymptotically among the n vehicles if $\xi_i(t) \to \xi_j(t), \forall i \neq j$, as $t \to \infty$ for all $\xi_i(0)$. The idea is to drive the states of all the n vehicles together through neighbor information sharing strategies. The Laplacian matrix or the communication matrix has to be a spanning tree to ensure the consensus [1].

The basic control strategy (6.3) can be extended for different formation control tasks. Instead of ensuring the convergence of $\xi_i(t) - \xi_j(t)$ to zero, an offset can be added here to achieve user specified formations. For example, a fixed offset δ on the x-axis and the zero offset on the y-axis can lead to the final axial alignment in a 2D space. The following algorithm is derived based upon (6.3):

$$u_i = \dot{\delta}_i - \sum_{j \in \mathcal{J}_i(t)} k_{ij}(t)[(\xi_i - \xi_j) - (\delta_i - \delta_j)], \quad i = 1, \ldots, n, \quad (6.4)$$

where $\delta_i - \delta_j$, $\forall i \neq j$ represents the user specified separation [6]. The different choices of δ_ℓ, $\ell = 1, \ldots, n$ can lead to different formation shapes.

6.1.2 Implementation of Consensus Algorithms

The above consensus algorithms (6.2) and (6.4) are for scenarios that the control input is the first-order derivative of the system state. There are also cases that the system state can be controlled directly. Similar consensus algorithms can be derived as follows. Suppose the unmanned vehicles can only move in a 2D space. Let $r_i = [x_i, y_i]^T$ and $r_i^d = [x_i^d, y_i^d]^T$ denote, respectively, the actual and desired position of vehicle i. For rendezvous, the control law is designed as

$$\dot{r}_i^d = - \sum_{j \in \mathcal{J}_i(t)} (r_i^d - r_j^d). \quad (6.5)$$

Another strategy is designed as

$$\dot{r}_i^d = - \sum_{j \in \mathcal{J}_i(t)} (r_i - r_j). \quad (6.6)$$

Both control laws could guarantee the $r_i^d(t) \to r_j^d(t)$ and $r_i(t) \to r_j(t)$, $\forall i \neq j$, asymptotically as $t \to \infty$ [6].

For axial alignment, the following algorithm is applied to update r_i^d as

$$\dot{r}_i^d = - \sum_{j \in \mathcal{J}_i(t)} [(r_i - r_j) - (\delta_i - \delta_j)], \quad (6.7)$$

where $\delta_i = [\delta_{ix}, \delta_{iy}]^T$ could be specified for specific missions based on vehicle sizes.

6.1.3 MASnet Hardware Platform

The Mobile Actuator and Sensor Network (MASnet) research platform in the Center for Self-Organizing and Intelligent Systems at Utah State University combines wireless sensor networks with mobility [7]. In other words, the robots can serve as both actuators and sensors. Although each robot has limited sensing, computation, and communication ability, they can coordinate with each other as a team to achieve

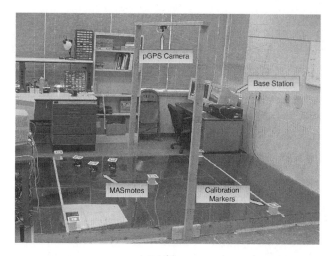

(a)

(b)

Figure 6.1 MASnet experimental platform. (a) MASnet testbed. (b) MASmote robot hardware.

challenging cooperative control tasks such as formation keeping and environment monitoring.

The MASnet platform comprises MASmotes, an overhead USB camera, and a base station PC, shown in Fig. 6.1. MASmotes are actually two-wheel differentially steered robots that can carry sensors and actuators with MicaZ from Crossbow serving as the processor. Thus, MASmotes support inter-motes communication, sensor data collecting, pulse width modulation signal generation. The overhead camera is used to determine each robot's position and orientation, called pseudo-GPS. Images are processed on the base PC, which also serves for serial to programming board communication and decision making. The base station communicates with a gateway mote mounted on a programming board by serial link. The gateway mote then communicates with the MASmotes over a 2.4-GHz wireless mesh network. Through communication the base station can send commands and pseudo-GPS information to each MASmote. All the MASmotes can also communicate with each other over the 2.4-GHz wireless mesh network.

The MASmote has a low-level PID controller to achieve an accurate position control, that is, r_i tracks r_i^d. So the specific control input for our MASmote are desired positions of robots: r_i^d. The consensus algorithm is implemented as below:

$$r_i^d = \frac{\sum_{j=1}^N g_{ij} r_j}{\sum_{j=1}^N g_{ij}},$$ (6.8)

where g_{ij} is the element from the communication matrix.

6.1.4 Experimental Results

Several consensus algorithms have been tested on the MASnet hardware simulation platforms including rendezvous, axial alignment, and V formation all with different communication topologies.

6.1.4.1 Rendezvous

Rendezvous is the basic for the consensus-based formation control. The key point is to achieve the convergence through constant communications. However, different communication topologies may affect the final convergence time and location. Three typical communication topologies were tested including the neighbor case (Case I), the isolated case (Case II), and the directed communication case (Case III), shown in Fig. 6.2. From the experimental results shown in Fig. 6.3, it can be seen that Case I achieves the rendezvous in the shortest time (11.58 s) because of more information sharing while Case II take a little longer time (18.89 s) and Case III can only achieve the local convergence.

Besides the fixed communication topologies, the switching ones were also tested on our hardware platform. One typical example is to compare one connected topology with its partial members. Communication topology $\mathcal{G}_u$ is decomposed into five partial

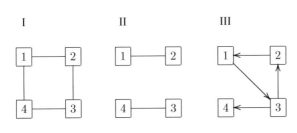

Figure 6.2 Communication topologies for the rendezvous experiment.

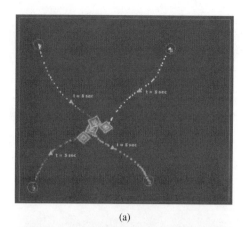

(a)

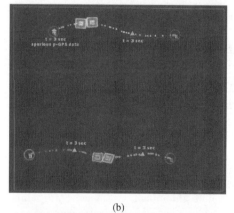

(b)

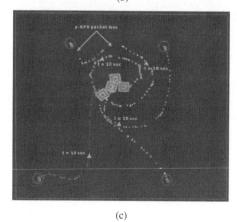

(c)

Figure 6.3 Experimental results of rendezvous for Cases I–III. (a) Case I ($t_f = 11.58$ s). (b) Case II ($t_f = 6.93$ s). (c) Case III ($t_f = 18.89$ s).

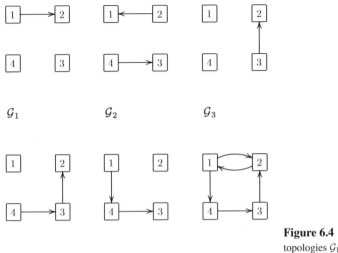

Figure 6.4 Switching information topologies $\mathcal{G}_1$–$\mathcal{G}_5$ and their union $\mathcal{G}_u$ for rendezvous.

members $\mathcal{G}_1$–$\mathcal{G}_5$, as shown in Fig. 6.4. The robot trajectories are shown in Fig. 6.5. It can be observed that both cases could arrive the final rendezvous, but they will perform differently while converging.

6.1.4.2 Axial Alignment

Axial alignment is essentially alignment along one axis of interest and in the two-dimensional case it can be demonstrated by aligning along one dimension. In order to avoid collisions in this process, the experiment involved alignment along the y-axis

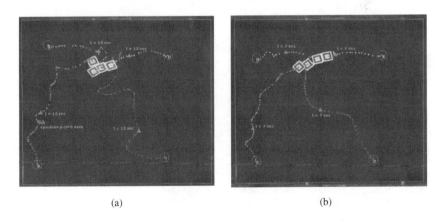

(a)   (b)

Figure 6.5 Experimental results of rendezvous with topologies randomly switched from $\bar{\mathcal{G}}_s$ versus a time-invariant topology $\mathcal{G}_u$. (a) Rendezvous with topologies randomly switching from $\bar{\mathcal{G}}_s = \{\mathcal{G}_1, \ldots, \mathcal{G}_5\}$ ($t_f = 45$ s). (b) Rendezvous with a time-invariant topology $\mathcal{G}_u$ ($t_f = 24$ s).

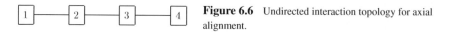

Figure 6.6   Undirected interaction topology for axial alignment.

with an offset alignment along the *x*-axis. By giving an desired offset from the center to the robots on the ends of the group, the line is spread evenly with spaces twice the distance of the offset. This simple mechanism achieves a relatively simple method of collision avoidance and also shows another attribute of the consensus algorithm. The communication topology is shown in Fig. 6.6 and the robot trajectories are shown in Fig. 6.7. The desired offset from the center for the outside robots is 20 cm, which gives a separation distance of 40 cm between all robots in the group along the *x*-axis. It can be observed that the robots can achieve the axial alignment mission with the consensus algorithm.

6.1.4.3   V Formation

The "V" formation control employs an extension of the offset idea. By setting an offset to the following members of the group, the group stays in a "V" formation with proper spacing. The spacing desired here, like the axial alignment experiment, is 40 cm in the *y* axis and also 40 cm in the *x* axis. This gives a separation of about 56.5 cm along the edge of the formation between any two robots. The communication flows in a leader–follower structure down through the group, shown in Fig. 6.8. The robot trajectories are shown in Fig. 6.9.

Due to the inherent limitations of the robots and the packet loss during vision problems from the pseudo GPS updates, formation moves on the MASnet platform are difficult. However, the formation is achieved through the move and stays in a general "V" shape throughout the move. This shows the robustness of the consensus

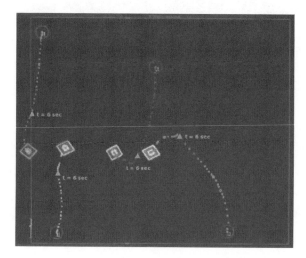

Figure 6.7   Experimental result of axial alignment.

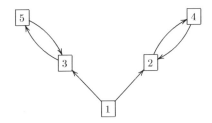

Figure 6.8 Information exchange topology for formation maneuvering.

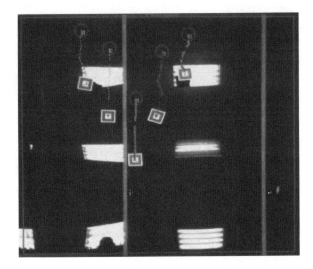

Figure 6.9 Experimental result of formation maneuvering.

algorithm and its ability to keep the group goal achieved even when motion as a particular formation is required.

6.2   SURFACE WIND PROFILE MEASUREMENT USING MULTIPLE UAVs

Winds in the atmospheric boundary layer (ABL) play an important role in human life because their flow cause the movements of heat, vapor, pollutants, and other airborne particles. Especially, processes such as forest fire spreading [8], gas leaking, pollution dispersion, or heat/vapor transfers are highly affected by the distribution of wind field. Surface winds are defined as flows less than 1 km above the ground. They could also be used to generate electricity [9]. To fully use the wind energy and accurately predict the dynamic changes of such diffusion/transfer processes, high-resolution and real-time wind field measurements are demanded. Traditional wind profiling techniques such as balloons, meteorological towers are not only expensive but also inflexible. In particular, for circumstances such as forest fires, it is almost impossible to have an accurate prediction for the local wind field based on data from weather stations due to its immobility and geographical sparse distribution [10]. Because of both temporal and spatial variations, it is hard to measure the wind field precisely in the meso- or micro-scale.

UAVs could be utilized as mobile platforms capable of carrying sensors in the 3D space [11]. Nowadays, technological advances in wireless networks and microelectromechanical systems make it possible to mount small, light weight, and inexpensive sensors on UAVs to measure the pressure, temperature, humidity, and wind speed in real time [12]. The advantages of using UAVs to measure the wind field include [13] the following:

- *Real Time*: UAVs provide real-time data to time critical situations such as pollutant diffusion, forest fire forecasting, combat urban environmental prediction, or chemical leaking.

- *High Resolution*: UAVs could get high-resolution microscale data without relying on the weather station. UAVs can be employed in all kinds of terrains.

- *Low Cost*: The developments of small UAVs make it affordable for its applications in meteorology.

- *More Flexibility*: UAVs could be easier to use compared with manned aircraft measurements since no human pilots will be endangered for such tedious measurement works.

- *More Data*: Multiple-UAV solution instead of only one. Easy for gradient-based searching and gradient information acquisition.

Many researchers have already looked into the problem of using multiple robots for the environmental monitoring problem. A model-free robot path planning algorithm is introduced for the coverage control problem of a density field [14]. The adaptive and singular value-based sampling algorithms are proposed for the ocean field sampling problem [15,16]. However, most of the reported efforts focus on the density field instead of a more complex vector field. Rotary-wing UAVs are also used for heat flux estimation with user customized pressure sensing unit [17]. To achieve the optimal measurement of the wind field, groups of UAVs could fly in formations for faster estimation [18].

In this book, small UAVs are defined as UAVs that weigh less than 55 lb [19]. They are mostly used for intelligence, surveillance, and reconnaissance missions in the troposphere, which is the part of the atmosphere about 15 km above the ground [20]. The lowest part of the troposphere is planetary boundary layer (PBL). The winds in the PBL, also called the surface winds, are highly affected by the Earth's surface and in turn play important roles in weather predictions. Researchers have used aircrafts to aid the measurements in the meteorology [21]. However, UAVs are still seldomly used for such meteorology missions. Typical small UAV platforms are able to fly in winds less than 20 m/s [11,22,23]. After the specific design and tests, small UAVs can even be deployed for very dangerous tornado tracking missions since no human pilots are required onboard.

Small UAVs can make real-time measurements of the wind, temperature, pressure, and humidity in points in a 3D space of the troposphere. The wind information is actually estimated based on the measurements of the relative airspeed and the absolute GPS speed of small UAVs. For instance, the horizontal wind estimation error of Kestrel autopilot can be as small as 5% on speed and 2% on heading [24]. There are also special pressure devices made for small rotary-wing UAVs to measure the vertical wind speed [17]. This information could be used to estimate the heat flux in the agricultural fields. There are a lot of potential wind applications for small UAVs including real-time wind profile mapping, forest fire prediction, fog diffusion prediction, and the like.

6.2.1 Problem Definition: Wind Profile Measurement

For common weather prediction problems, the synoptic wind field in the horizon order of 1000 km is considered. However, UAVs are more suitable for mesoscale or microscale wind field measurement. The mesoscale field is defined with the horizon length between several kilometers to several hundred kilometers while the microscale is as small as 1 km or less. The 2D meso- or microscale wind field measurement problem is focused in this chapter. Assuming a near-surface wind field of interest by $V = (v_e, v_n)^T$, the vorticity ξ and divergence δ are defined as follows [25,26]:

$$\xi = k \cdot \nabla \times V, \tag{6.9}$$

$$\delta = \nabla \cdot V, \tag{6.10}$$

where k is the unit vector in z axis.

Without loss of generality, V could be decomposed into two vectors: a nondivergent one and a curl-free one.

$$V = k \times \nabla \psi + \nabla \chi, \tag{6.11}$$

where

$$\begin{cases} \nabla \cdot (k \times \nabla \psi) = 0, \\ \nabla \times (\nabla \chi) = 0. \end{cases}$$

We have the following assumptions to guarantee a unique solution for the above equations [25]:

- At the boundary, χ is assumed to be $\chi = 0$.
- At the boundary, p is assumed to be bounded and have higher order derivatives.
- ψ is assumed to be linear with respect to the pressure field in the area of interest.
- Assume that the ξ and δ could be approximated by low-order polynomials.

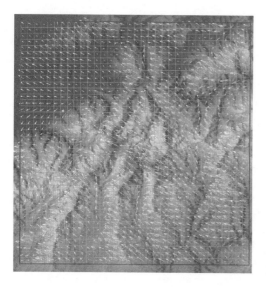

Figure 6.10   A random wind field.

The Poisson equations could be derived from (6.11)

$$\nabla^2 \psi = \xi, \tag{6.12}$$

$$\nabla^2 \chi = \delta, \tag{6.13}$$

$$v_e = -\frac{\partial p}{\partial y} + \frac{\partial \chi}{\partial x}, \tag{6.14}$$

$$v_n = \frac{\partial p}{\partial x} + \frac{\partial \chi}{\partial y}, \tag{6.15}$$

where ξ and δ have the following forms:

$$\xi(x, y) = \sum_{m=0}^{M_c} \sum_{n=0}^{M_c} c_{m,n} x^m y^n \quad \text{for} \quad m + n \le M_c, \tag{6.16}$$

$$\delta(x, y) = \sum_{m=0}^{M_c} \sum_{n=0}^{M_c} d_{m,n} x^m y^n \quad \text{for} \quad m + n \le M_c. \tag{6.17}$$

A typical meso-/microscale wind field (about 10×10 km) is shown in Fig. 6.10 with 280 m resolution at 100 m high. The example wind field is generated by Wind-Station simulation software [27]. It can be seen in Fig. 6.10 that the wind field is greatly affected by the complex mountain terrain.

6.2.1.1   Numerical Solution for Parameter Identification

Assuming that the wind field in the whole area of interest can be measured, a numerical solution can then be derived [28]. Given an area with $M \times N$ equally spaced grids with spacing h, the measurements of the velocity at all the points form two vectors $(VN_{MN \times 1} \; VE_{MN \times 1})$.

$$\begin{bmatrix} VN \\ VE \end{bmatrix} = F\hat{W}, \qquad (6.18)$$

where $\hat{W}$ is the vector containing all the unknown parameters and pressure field boundary conditions; F is determined from the corresponding governing difference equations. Then,

$$\hat{W} = (F^T F)^{-1} F^T \begin{bmatrix} VN \\ VE \end{bmatrix}. \qquad (6.19)$$

The objective function of the wind profile measurement could be defined as

$$\min \; \sum_{i=1}^{MN} [(V_i - \hat{V}_i)(V_i - \hat{V}_i)^T], \qquad (6.20)$$

where $\hat{V}$ is the estimate of the wind field from the identified parameters $\hat{W}$.

6.2.2   Wind Profile Measurement Using UAVs

UAV could serve as a new measurement platform for the surface wind field because UAVs could be easily commanded to any 3D point in the ABL, which is a hard mission for meteorological stations. Common microsensors such as temperature, humidity, pressure, and wind speed are now very cheap to be mounted on the UAVs. Recently, it is even possible to employ multiple small UAVs to form a team to measure the horizontal or vertical profile simultaneously, which can offer big advantages for later gradient detections.

The most intuitive way of using a single UAV for wind profile measurement is to map the wind field first and then analyze the model off-line. Assume that the area of interest is rectangular ($utm_e \in [x_1, x_2]$, $utm_n \in [y_1, y_2]$), the desired resolution is h m. The UAV is assumed to be able to fly at a certain preplanned altitude with the cruising speed v_c and a minimal turning radius $R_{min} < h$. In other words, the UAV could follow either a straight line or a circle with radius greater than R_{min}. Sensor readings including the temperature, pressure, wind speed, and GPS position are sent to the ground station at f Hz in real time. The UAV trajectory could be planned following the algorithm described in Chapter 5. The sweep distance is h and the sensor data collected while turning is neglected for the consistency in spatial resolution of the collected data.

The similar numerical method could be used for the wind field parameter identifications after collecting all the wind data. However, this method is time consuming and especially not efficient for relatively simple wind fields. In fact, the model identification accuracy could be used as a feedback to guide the UAV to an information richer area instead of following the preplanned trajectories without knowing the specifics of the local wind field.

The objective function of the wind profile measurement incrementally using UAVs could be defined as

$$\min \quad E[(V_k - \hat{V}_k)(V_k - \hat{V}_k)^T], \tag{6.21}$$

where V_k is the real wind measurement at time k and $\hat{V}_k$ is the estimate of the wind field based on the historic data from $t = 0, \cdots, k$.

The difference of the above problem with the numerical off-line solution is that the UAV needs to be guided based on the estimation errors. To solve this problem, we propose the following scheme.

6.2.2.1 Initial Estimation of Prevailing Wind

The prevailing wind is also called longitudinal wind, which is the wind prediction estimated from larger scale such as synoptic scale. It is the prevailing direction at the predicted height without considering the surface frictions and complex terrains. The idea here is to send the UAV to measure the wind speed and estimate the noise level at the same time. In this way, the UAV does not have to visit all the grids.

Assume the wind field could be modeled as

$$u = \bar{u} + \xi_u,$$
$$v = \bar{v} + \xi_v, \tag{6.22}$$

where $\bar{u} + \bar{v}$ is called the constant longitudinal wind and $\xi_u \sim N(0, \sigma_u)$ and $\xi_v \sim N(0, \sigma_v)$ are white noises. The parameters of the prevailing wind model could be identified by Kalman filter described as below:

$$vx_k^- = vx_{k-1} + \xi_{vx},$$
$$vy_k^- = vy_{k-1} + \xi_{vy}, \tag{6.23}$$

$$vx_k = vx_k^-,$$
$$vy_k = vy_k^-. \tag{6.24}$$

6.2.2.2 Recursive Parameter Estimation Using UAVs

With the real-time sensor data coming in, the local wind field model can be derived. The estimated model is then compared for each small division. The UAV will be sent to the area with the biggest difference compared with the identified model (the gradient direction).

$$\begin{bmatrix} VN \\ VE \end{bmatrix} = \nabla e. \tag{6.25}$$

6.2.2.3   Optimal Wind Profile Measurement Procedures

The path planning algorithm using a single UAV for optimal wind profile measurement can be described as follows:

(1) Launch the UAV and send it into the desired area and altitude for survey.

(2) Using Kalman filter to make an initial guess of the wind model, or the prevailing wind speed $\bar{u}$ and covariances.

(3) If $\mathrm{mod}(k, l) = 0$, estimate the wind field model based upon the collected wind data, where l is the preset time interval to trigger the wind model prediction.

(4) If $\sum_{i=k}^{k+l} |E_i - \hat{E}_i| > \sigma l$, go to Step 5. Otherwise, go to Step 6. σ is the max tolerance of the model errors.

(5) Follow the flight direction from (6.25), go back to Step 3.

(6) Follow the preplanned flight path, go back to Step 3.

6.2.3   Wind Profile Measurement Using Multiple UAVs

The intuitive way of using multiple UAVs in wind profile measurement is to divide the interested domain into small regions for every single UAV. However, one advantage here of using UAV formation flight is the real-time estimation of the pressure derivative $\left(\frac{\partial p}{\partial x}, \frac{\partial p}{\partial y}\right)$. Assuming that four UAVs collect data while maintaining a simplex shape with h distances, $\frac{\partial p}{\partial x}$, $\frac{\partial p}{\partial y}$, and $\frac{\partial p}{\partial z}$ can then be measured in real time, which in turn can reduce the computation time for the later parameter estimations. The simplex formation is shown in Fig. 6.11.

The formation shape could also be changed based on different resolution requirements. For example, a closer formation is needed for areas with turbulence while a

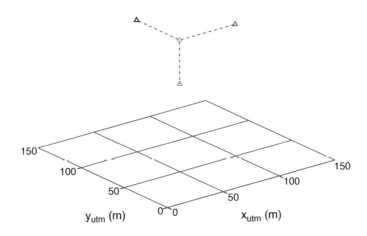

Figure 6.11   UAVs flying in a simplex formation.

formation of larger separation is chosen for prevailing wind dominant areas. The UAV distance could also be determined by the resolution requirements.

6.2.4 Preliminary Simulation and Experimental Results

The preliminary simulation and experimental results are provided in this section, focusing on the simple wind field model.

6.2.4.1 Simulation Results

To simplify the problem, a simple wind field is considered first, which can be modeled by (6.22). This model is generally true for most open areas with no sudden wind changes in several hours. The target wind field is a rectangular area about 3 km × 3 km. The nominal wind speed is 5 m/s and the nominal wind direction is 30 degrees (relative to the north). The UAV wind estimations are corrupted with the measurement noise $\xi_u \sim N(0, 1)$ and $\xi_v \sim N(0, 1)$, shown in Fig. 6.12.

A single UAV is sent out for this mission following trajectories planned based on the method described in Chapter 5. The UAV is assumed to fly at 15 m/s and report wind measurements every 3.3 s to the ground control station (GCS). The proposed Kalman filter algorithm is running on the GCS for the real-time wind field estimation. The error covariances are estimated through real measurements. The estimation results of the east wind ($\bar{u}$) and north wind ($\bar{v}$) are shown in Figs. 6.13 and 6.14, respectively. It can be observed that the nominal wind can be accurately estimated after several minutes duration of flight.

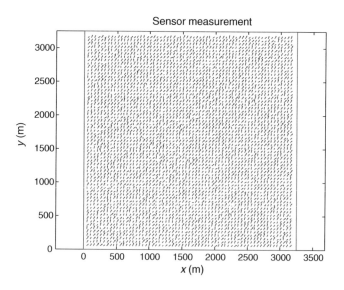

Figure 6.12   Wind measurement (simulation).

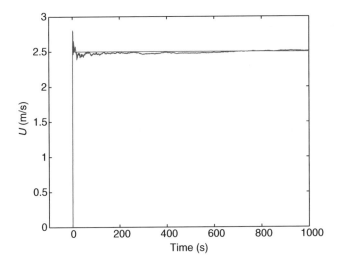

Figure 6.13 Estimation of wind speed $\bar{u}$.

The nominal wind estimation algorithm can also be used as an initial guess for a more complex wind field. The UAV could be sent to fly at the flat area first to get an estimation of the prevailing wind and then be deployed to complex terrains for further wind model estimation, which is left as future work.

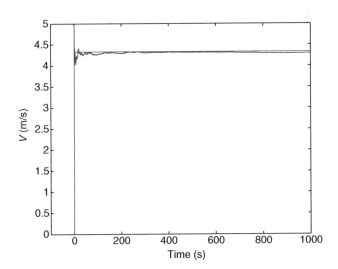

Figure 6.14 Estimation of wind speed $\bar{v}$.

Table 6.1   AggieAir UAS Specifications

	Specifications
Flight time	≤ 1 h
Cruise speed	15–20 m/s
Flight altitude	≤ 1000 m (AGL)
Available sensors	Temperature/pressure/air speed
Navigation sensors	GPS/IMU

6.2.4.2 Experimental Results

The wind estimation data collected during the flight of a single UAV is used in this section to test the nominal wind estimation algorithm.

Tiger UAV Platform Introduction   AggieAir-Tiger UAV is used as the experimental platform for the wind data collection [29]. It is a flying-wing UAV with a 72-in wingspan called Tiger, with the image shown in Chapter 2. The weight of Tiger UAV is about 8 pounds including batteries, GPS, modem, autopilot, motors, and sensors. Open-source Paparazzi TWOG autopilot board is used as the autopilot hardware [30] with Microstrain GX2 IMU and u-blox LEA-5H GPS unit for inertial measurement. The specifications of AggieAir-Tiger UAV are described in Table 6.1. All the sensor readings are sent to the GCS through a 900-MHz serial modem in real time. The wind is estimated on the GCS with the UAV flight trajectory, the GPS speed, and the throttle percentage [30].

Preliminary Results   The flight data were collected at the Cache Junction research farm of Utah State University from 9:54 to 10:27 AM on May 31, 2009. The Tiger UAV was used to collect data at 90 m above the ground. The UAV flight trajectory and the estimated wind data are shown in Fig. 6.15. The UAV was first sent into a circle to gain the specified altitude. Then, it started to follow the survey of a rectangular field. The cruise speed is set as 15 m/s and the wind is estimated every 10 s.

Since the research farm is relatively flat, the prevailing wind can be estimated using the proposed algorithms, shown in Figs. 6.16 and 6.17. It can be seen that the north wind estimation converges after 10 min flight, shown in Fig. 6.17. Unfortunately, there are no other wind measurement techniques available for validations. One possible solution is to use multiple UAVs to see if they could achieve a consensus over the wind data.

Future Work   The simulation and experimental results in this chapter mainly focus on the estimation of the prevailing wind. It is only the first step to accurately model a microscale wind field. More efforts are needed for a more general wind field estimation problem including

- More accurate wind estimation techniques. The current estimation algorithm for Paparazzi UAV platform is not accurate and is not calibrated. Further efforts are needed for accurate wind estimation using small UAVs.

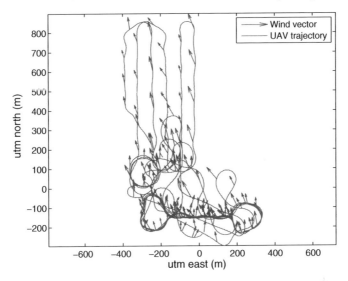

Figure 6.15   Example wind field measured by AggieAir UAV.

- More wind information to measure. Except for horizontal winds, vertical winds are also very important especially to the heat, vapor transportation problems. Air speed and pressure sensors can also be installed on vertical take-off and landing (VTOL) UAVs for more wind information following specific flight patterns;
- Further theoretical analysis for the complex wind field model. The wind fields in the complex domain such as mountain and urban areas are highly dynamic

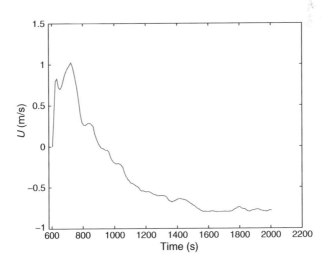

Figure 6.16   Estimation of wind speed U.

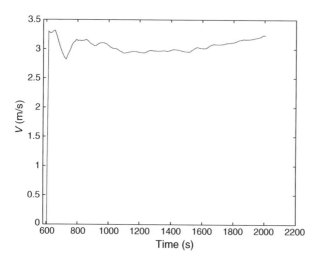

Figure 6.17 Estimation of wind speed V.

and hard to measure. However, wind fields in these areas could have huge impacts on the problems such as fire evolution. A more general wind model is indispensable for such missions.

6.3 CHAPTER SUMMARY

The remote sensing problem using multiple UAVs is looked into in this chapter. First, the multivehicle consensus algorithm is tested on the indoor MASnet robotics platform with different communication topologies and different missions. Then the wind profiling measurement problem is focused. A new path planning algorithm for wind field measurement is proposed using single or multiple UAVs. Preliminary simulation and flight results showed the potential of the algorithm.

REFERENCES

1. W. Ren and R. W. Beard. Consensus seeking in multiagent systems under dynamically changing interaction topologies. *IEEE Transactions on Automatic Control*, 50(5): 655–661, 2005.

2. A. Jadbabaie, J. Lin, and A. S. Morse. Coordination of groups of mobile autonomous agents using nearest neighbor rules. *IEEE Transactions on Automatic Control*, 48(6):988–1001, 2003.

3. R. Olfati-Saber and R. M. Murray. Consensus problems in networks of agents with switching topology and time-delays. *IEEE Transactions on Automatic Control*, 49(9): 1520–1533, 2004.

4. J. A. Fax and R. M. Murray. Information flow and cooperative control of vehicle formations. *IEEE Transactions on Automatic Control*, 49(9):1465–1476, 2004.

5. Z. Lin, M. Broucke, and B. Francis. Local control strategies for groups of mobile autonomous agents. *IEEE Transactions on Automatic Control*, 49(4):622–629, 2004.

6. W. Ren, H. Chao, W. Bourgeous, N. Sorensen, and Y. Q. Chen. Experimental validation of consensus algorithms for multi-vehicle cooperative control. *IEEE Transactions on Control Systems Technology*, 16(4):745–752, 2008.

7. W. Bourgeous, L. Ma, P. Chen, Z. Song, and Y. Q. Chen. Simple and efficient extrinsic camera calibration based on a rational model. In *Proceedings of the IEEE Conference on Mechatronics and Automation*, pages 177–182, June 2006.

8. B. Butler, R. Stratton, J. Forthofer, M. Finney, L. Bradshaw, and D. Jimenez. Detailed wind information and at its application to improved firefighter safety. In *Proceedings of the Eighth International Wildland Fire Safety Summit*, 2005.

9. L. Y. Pao and K. E. Johnson. A tutorial on the dynamics and control of wind turbines and wind farms. In *Proceedings of the American Control Conference*, pages 2076–2089, June 2009.

10. J. M. Forthofer, B. W. Butler, K. S. Shannon, M. A. Finney, L. S. Bradshaw, and R. Stratton. Predicting surface winds in complex terrain for use in fire growth models. In *Proceedings of the 5th Symposium on Fire and Forest Meteorology and the 2nd International Wildland Fire Ecology and Fire Management Congress*, November 2003.

11. H. Chao, A. M. Jensen, Y. Han, Y. Q. Chen, and M. McKee. *Advances in Geoscience and Remote Sensing*, chapter AggieAir: towards low-cost cooperative multispectral remote sensing using small unmanned aircraft systems. IN-TECH, Vukovar, Croatia, 2009.

12. H. Chao, Y. Cao, and Y. Q. Chen. Autopilots for small unmanned aerial vehicles: a survey. *International Journal of Control, Automation, and Systems*, 8(1):36–44, 2010.

13. H. Chao, M. Baumann, A. M. Jensen, Y. Q. Chen, Y. Cao, W. Ren, and M. McKee. Band-reconfigurable multi-UAV-based cooperative remote sensing for real-time water management and distributed irrigation control. In *Proceedings of the International Federal of Automatic Control (IFAC) World Congress*, pages 11744–11749, July 2008.

14. J. Cortés, S. Matínez, T. Karatas, and F. Bullo. Coverage control for mobile sensing networks. *IEEE Transactions on Robotics and Automation*, 20(20):243–255, 2004.

15. A. Caiti, G. Casalino, E. Lorenzi, A. Turetta, and R. Viviani. Distributed adaptive environmental sampling with AUVs: cooperation and team coordination through minimum-spanning-tree graph searching algorithms. In *Proceedings of the Second IFAC Workshop on Navigation, Guidance and Control of Underwater Vehicles*, April 2008.

16. D. O. Popa, A. C. Sanderson, V. Hombal, R. J. Komerska, S. S. Mupparapu, D. R. Blidberg, and S. G. Chappel. Optimal sampling using singular value decomposition of the parameter variance space. In *Proceedings of the IEEE/RSJ International Conference on Intelligent Robots and Systems*, pages 3131–3136, August 2005.

17. J. Bange, P. Zittel, T. Spieß, J. Uhlenbrock, and F. Beyrich. A new method for the determination of area-averaged turbulent surface fluxes from low-level flights using inverse models. *Boundary Layer Meteorology*, 119(3):527–561, 2006.

18. Y. Q. Chen and Z. Wang. Formation control: a review and a new consideration. In *Proceedings of the IEEE/RSJ International Conference on Intelligent Robots and Systems*, pages 3181–3186, August 2005.

19. B. Tarbert, T. Wierzbanowski, E. Chernoff, and P. Egan. Comprehensive set of recommendations for sUAS regulatory development. Technical Report, Small UAS Aviation Rulemaking Committee, 2009.

20. J. M. Wallace and P. V. Hobbs. *Atmospheric Science: An Introductory Survey*. Academic Press, Orlando, FL, 1977.

21. D. H. Lenschowa, V. S. Jovcicb, and B. Stevens. Divergence and vorticity from aircraft air motion measurements. *Journal of Atmospheric and Oceanic Technology*, 24(12):2062–2072, 2007.

22. A. Rodriguez, E. Andersen, J. Bradley, and C. Taylor. Wind estimation using an optical flow sensor on a miniature air vehicle. In *Proceedings of the AIAA Guidance, Navigation and Control Conference*, number AIAA-2007-6614, August 2007.

23. H. J. Palanthandalam-Madapusi, A. Girard, and D. S. Bernstein. Wind-field reconstruction using flight data. In *Proceedings of the American Control Conference*, pages 1863–1868, June 2008.

24. Procerus Technologies. Kestrel autopilot specifications, 2008. http://www.procerusuav.com.

25. D. G. Long. Modelling and measurement of near-surface oceanic winds. *International Journal of Modelling and Simulation*, 13(4):156–161, 1993.

26. P. Lynch. Deducing the wind from vorticity and divergence. *Monthly Weather Review*, 116(1):86–93, 1987.

27. A. M. Gameiro Lopes. WindStation software, 2010. http://www.easycfd.net/windstation.

28. D. G. Long and J. M. Mendel. Model-based estimation of wind fields over the ocean from wind scatterometer measurements. i. development of the wind field model. *IEEE Transactions on Geoscience and Remote Sensing*, 28(3):349–360, 1990.

29. CSOIS. OSAM UAV website, 2008. http://www.engr.usu.edu/wiki/index.php/OSAM.

30. Open Source Paparazzi UAV Project, 2008. http://www.recherche.enac.fr/paparazzi/.

Chapter 7

Diffusion Control Using Mobile Sensor and Actuator Networks

7.1 MOTIVATION AND BACKGROUND

Diffusion processes such as chemical/radiation leaks, oil spills can have a large impact on human health and natural environment. Nowadays, technological advances in networking and micro eletro mechanical systems (MEMS) make it possible to employ a large number of mobile/static sensors/actuators to observe the diffusion, locate the source, and even counterreact with the harmful pollutants when a mobile sprayer network is used. In the past decade, many researchers looked into this topic. A swarm of mobile robots are used to detect chemical plume source with gradient climbing [1]; a moving diffusion source can be identified based on the parameter estimation algorithm [2]; boundary estimation and following problems are considered [3]. However, only the source information is not enough for controlling a diffusion process. Centroidal Voronoi tessellations (CVTs) are introduced in coverage control of a static gradient field with mobile sensor networks [4–6] and extended to a diffusing and spaying scenario [7].

The monitoring and control of a diffusion process can be viewed as an optimal sensor/actuator placement problem in a distributed system [8]. Basically, a series of desired actuator positions are generated based on centroidal Voronoi tessellations and integrated with PID controllers for neutralizing control based on Voronoi partitions. CVT algorithm provides a non-model-based method for coverage control and diffusion control using groups of vehicles. The CVT algorithm is robust and scalable [9,10], and it can guarantee the groups asymptotically converging to the affected area even in multiple/mobile sources application [4].

Consensus is a common agreement reached by a group as a whole. The consensus can be made on robot formation, source location tracking, task assignment, and traffic control [11–13]. Although a group of mobile actuators are used for the

Remote Sensing and Actuation Using Unmanned Vehicles, First Edition. Haiyang Chao and YangQuan Chen.
© 2012 by The Institute of Electrical and Electronics Engineers, Inc.
Published 2012 by John Wiley & Sons, Inc.

diffusion control [7], the communication and information aspects are not taken care of. The mobile actuator only negotiates with its neighboring sensors, not neighboring actuators/sprayers, on how much to spray and where to go. As will be shown in this chapter, the information sharing and interaction among neighboring actuators/sprayers in a group can have a large impact on the coordinated movements of these actuators and the resulted control performance consequently. Since the actuators are sent out for the same task, consensus is needed on both where to spray and how much to spray. The mobile actuators need to get close to the polluted area but it is not efficient to cluster, or running together densely. On the other hand, the neutralizer spraying should also be balanced since the best energy saving way is to maximize the neutralizing ability of every actuator. A new consensus algorithm is introduced and integrated into the CVT algorithm to guarantee the actuator group to converge faster toward the affected area with an improved control performance.

This chapter is organized as follows. In Section 7.2, the diffusion process is modeled by a partial differential equation (PDE) equation and the diffusion control problem is formulated. In Section 7.3, CVT-based optimal actuator location algorithm is briefly introduced. Section 7.4 focuses on the grouping effect while using different actuator group sizes for CVT-based diffusion control problem. Section 7.5 is devoted to introducing the information consensus into the CVT-based optimal actuator location algorithm. Finally, simulation results and comparisons with the plain CVT algorithm are presented in Section 7.6.

7.2 MATHEMATICAL MODELING AND PROBLEM FORMULATION

In this section, the PDE mathematical model of a diffusion process is introduced and the neutralizing control problem is then formulated.

Suppose a diffusion process evolves in a convex polytope Ω: $\Omega \in \mathcal{R}^2$. $\rho(x, y)$: $\Omega \rightarrow \mathcal{R}_+$ is used to represent the pollutant concentration over Ω. The dynamic process can be modeled with the following partial differential equation (PDE):

$$\frac{\partial \rho}{\partial t} = k \left(\frac{\partial^2 \rho}{\partial x^2} + \frac{\partial^2 \rho}{\partial y^2} \right) + f_d(x, y, t) + f_c(\tilde{\rho}, x, y, t), \tag{7.1}$$

where k is a positive constant representing the diffusing rate; $f_d(x, y, t)$ shows the pollution source; $\tilde{\rho}$ is the measured sensor data; and $f_c(\tilde{\rho}, x, y, t)$ is the control input applied to the system, which represents the effect of neutralizing chemicals sent out by mobile actuators to counteract the pollutants. An example PDE system is shown in Fig. 7.1.

Assume n mobile actuators are sent to the field $f_c = f_{c_1} + \cdots + f_{c_n}$. $P = (p_1, \cdots, p_n)$ represent the locations of n actuators, $| \cdot |$ is the Euclidean distance. n actuators partition Ω into a collection of n Voronoi diagrams $\mathcal{V} = \{V_1, \cdots, V_n\}$, $p_i \in V_i$, $V_i \cap V_j = \emptyset$ for $i \neq j$.

$$V_i = \{q \in \Omega \mid |q - z_i| < |q - z_j| \text{ for } j = 1, \cdots, n, j \neq i\} \tag{7.2}$$

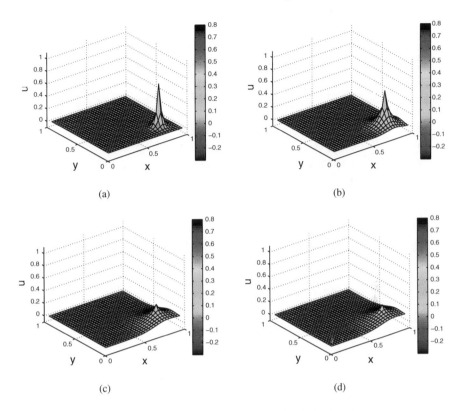

(a)   (b)

(c)   (d)

Figure 7.1   Surf plot of a diffusion process modeled by (7.1): $k = 0.01$, $f_c = 0$, $f_d = f_d(0.8, 0.2, 20e^{-t})$. (a) $t = 0.2s$. (b) $t = 0.8s$. (c) $t = 2s$. (d) $t = 2.2s$.

The control objectives are

- to control the diffusion of the pollution to a limited area
- to neutralize the pollution as quickly as possible without making the area of interest overdosed.

To achieve the above requirements, the following evaluation equation needs to be minimized [4,7]:

$$\min. \mathcal{K}(P, \mathcal{V}) = \sum_{i=1}^{n} \int_{V_i} \rho(q) |q - p_i|^2 \mathrm{d}q \text{ for } q \in \Omega,$$

$$s.t. \quad |\dot{p}_i| < k_v, |\ddot{p}_i| < k_a, \sum_{i=1}^{n} \int u_{spray_i}(t) \mathrm{d}t < k_s, \tag{7.3}$$

where $\dot{p}_i$ and $\ddot{p}_i$ represent the first- and second-order dynamics of the actuator and $u_{spray_i}(t)$ is the neutralizing control input of the actuator i at time t.

Define the mass and centroid of region V_i as

$$M_{V_i} = \int_{V_i} \rho(q) dq,$$

$$\bar{p}_i = \frac{\int_{V_i} q\rho(q) dq}{\int_{V_i} \rho(q) dq}.$$

To minimize $\mathcal{K}$, the distance $|q - p_i|$ should be small when the pollution concentration $\rho(q)$ is high. But it is not an efficient strategy to drive all actuators very close to the pollution source, because the diffused pollutants far away from the source need also be neutralized quickly to minimize (7.3). A necessary condition to minimize $\mathcal{K}$ for coverage control in a static gradient field is that $\{p_i, V_i\}_{i=1}^n$ is a CVT of Ω [4].

$$\frac{\partial \mathcal{K}}{\partial p_i} = 2M_{V_i}(p_i - \bar{p}_i) \tag{7.4}$$

The CVT algorithm is further extended to a dynamical diffusion process [7]. It is based on a discrete version of (7.3) and the concentration information comes from the measurements of the static, low-cost mesh sensors. The diffusion control problem is converted to two subproblems: location optimization (where to go for actuators) and neutralizing control (how much to spray).

7.3 CVT-BASED DYNAMICAL ACTUATOR MOTION SCHEDULING ALGORITHM

In this section, CVT-based actuator motion planning algorithm is discussed in details.

The classic Lloyd's algorithm [6,14] is an iterative algorithm to generate a centroidal Voronoi diagram from any set of generating points. It is modified to achieve coverage control [4] and diffusion control [7].

7.3.1 Motion Planning for Actuators with the First-Order Dynamics

Assume that the sensors can be modeled by a first-order dynamical equation:

$$\dot{p}_i = u_i. \tag{7.5}$$

To minimize $\mathcal{K}$ in (7.3), the control input is set to be

$$u_i = -k_p(p_i - \bar{p}_i), \tag{7.6}$$

where k_p is a positive gain and $\bar{p}_i$ is the mass centroid of V_i. $\bar{p}_i$ is time-variant with diffusing.

7.3.2 Motion Planning for Actuators with the Second-Order Dynamics

If the second-order dynamical sensor model is used, similarly we have

$$\ddot{p}_i = u_i. \tag{7.7}$$

To minimize $\mathcal{K}$ in (7.3), the control input is set to be

$$u_i = -k_p(p_i - \bar{p}_i) - k_d \dot{p}_i, \tag{7.8}$$

where both k_p and k_d are positive constants.

The latter part of (7.8) $k_d \dot{p}_i$ is the viscous friction introduced [15], where k_d is the friction coefficient and $\dot{p}_i$ represents the velocity of the robot i. This part is used for eliminating the oscillatory behavior [16] when the robot gets close to its destination. The viscous term guarantees the robot coming to a standstill state even with no external force.

7.3.3 Neutralizing Control

Proportional control is used for the neutralizing chemical releasing. The amount of chemicals each robot releases is proportional to the average pollutant concentration in the Voronoi cell belonging to that robot.

$$u_{spray_i}(t) = -k_{pr} \frac{\int_{\bar{V}_i} \rho(x, y) dV}{\int_{\bar{V}_i} dV}, \tag{7.9}$$

$\bar{V}_i = V_i \cap C_i$ where $C_i = \{q | |q - p_i| < r_i\}$, r_i represents the sensing range of ith actuator, and V_i is the Voronoi diagram of actuator i.

7.4 GROUPING EFFECT IN CVT-BASED DIFFUSION CONTROL

In this section, we consider the distributed control of a time-varying pollution diffusion process using groups of mobile actuators with an emphasis on how different grouping methods affect the control performance. The scenario is described as follows: A toxic diffusion source is releasing toxic gas/fog in a 2D domain. The diffusion process is modeled as a PDE system and we assume static mesh sensor networks are deployed in the polluted area to measure chemical concentration. Then, a few mobile robots equipped with controllable dispensers of neutralizing chemicals are sent out to counteract the pollution by properly releasing the neutralizing chemicals. The remaining part of this section is organized as follows. We analyze the trade-off between group size and the efficiency in actuator control. The simulation results and comparisons are presented.

7.4.1 Grouping for CVT-Based Diffusion Control

In this section, we discuss in detail if CVT-based algorithm could be used to large numbers of actuator groups and how to decide the appropriate grouping size according to the final performance requirements. In [7], we have shown that the CVT algorithm works well for four mobile actuators. It is obvious that we can achieve better control result for the pollution neutralization by using more mobile actuators. But a large group size also has trade-offs such as more computation and communication requirements, which lower the efficiency and robustness of control system.

With the increasing of actuator numbers, we need to use the computational complexity theory to test if the CVT will impose a large computation burden. There are many practical methods for constructing Voronoi diagrams including the naive method [14], the flip method, and the incremental method. Specifically, we chose the Delaunay triangulation method based on Qhull. According to [17], the computational complexity is

$$f_r = O\left(\frac{r^{d/2}}{(d/2)!}\right),$$

where d is the dimension; n the number of input points, r the number of processed points, and f_r the maximum number of facets of r vertices. For our problem, $d = 2$.

$$f_r = O(r).$$

For simulation purpose, we use `delaunay()` and `voronoi()` functions in MATLAB to get the CVT diagrams. We found no large computation burdens.

Next, we will show how the CVT algorithm can be implemented in a distributed way. That is, the algorithm can be executed on a group of robots instead of a centralized one. In fact, we need only get p_i and $\overline{p_i} = C_{V_i}$ for every time step. To get a distributed implementation, each actuator needs to know the relative location of each Voronoi neighbor for computing its own Voronoi cell.

Given the above discussion on computational cost, it is feasible to consider more actuators for pollution neutralizing problem. However, we are interested in how many actuators in a subgroup or what the best grouping size is. In real circumstances, for a fair comparison, we must use the same amount of neutralizing chemicals for various numbers of groups. There should be an optimal group size given a specific performance metric.

7.4.2 Diffusion Control Simulation with Different Group Sizes

`Diff-MAS2D` is used as the simulation platform for our implementation. The area concerned is given by $\Omega = \{(x, y)|0 \le x \le 1, 0 \le y \le 1\}$. The system with control input is modeled as

$$\frac{\partial \rho(x, y, t)}{\partial t} = k(\frac{\partial^2 \rho(x, y, t)}{\partial x^2} + \frac{\partial^2 \rho(x, y, t)}{\partial y^2}) + f_c(x, y, t)$$

$$+ f_d(x, y, t), \tag{7.10}$$

where $k = 0.01$ and the Neumann boundary condition is given by

$$\frac{\partial u}{\partial n} = 0.$$

where n is the outward direction normal to the boundary.

The stationary pollution source is modeled as a point disturbance f_d to the PDE system (7.10) with its position at $(0.75, 0.35)$ and

$$f_d(t) = 20e^{-t}|_{(x=0.75, y=0.35)}.$$

In our simulation, we assume that once deployed, the mesh sensors remain static. There are 29×29 sensors evenly distributed in a square area $(0, 1)^2$ and they form a mesh over the area. There are four mobile robots that can release the neutralizing chemicals. For the robot motion control, the viscous coefficient is given by $k_v = 1$ and the control input is given by

$$F_i = -3(p_i - \bar{p}_i) - \dot{p}_i.$$

The pollution source begins to diffuse at $t = 0$ to the area Ω and initially the mobile actuator robots are evenly distributed within the domain Ω (one by one square) at the following specific positions: $(0.5, 0.5)$ for $1 * 1$ grouping case; for $2 * 2$ grouping case, $(0.33, 0.33)$, $(0.33, 0.66)$, $(0.66, 0.33)$, $(0.66, 0.66)$, respectively, and so on and so forth.

We choose the simulation time to $t = 5$ s and the time step as $\Delta t = 0.002$ s. The robot recomputes its desired position every 0.2 s. To show how the robots can control the diffusion of the pollutants, the robots begin to react at $t = 0.4$ s. The system evolves under the effects of diffusion of pollutants and diffusion of neutralizing chemicals released by robots. To show the scalability of the CVT algorithm for bigger groups, the control results are shown by using 5*5 and 9*9 mobile actuators respectively. Table 7.1 shows the computation time on the PC (P4-2.6G, 256M RAM) and the remaining pollutants at the end of the simulation. It can be seen that the computational load does not increase much with the increase of the number of actuator groups. In Fig. 7.2, the y-axis is the sum of the mesh sensor measurements.

Table 7.1 Computational Time for Simulation and Control Results

Grouping	Time for simulation (s)	Remaining pollutants
2*2	510	3.3994
3*3	582	0.7046
4*4	615	0.3372

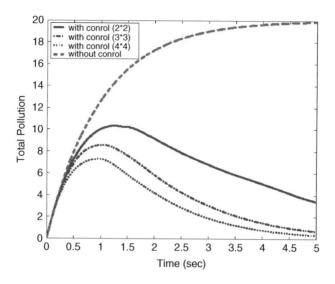

Figure 7.2   Evolution of the amount of pollutants (2*2, 3*3 and 4*4 robots).

To compare the performance of different groupings of actuators, we compare our CVT-based algorithm with the uniformly distributed case. The control laws for chemical releasing are the same. But Table 7.2 shows the different output parameter for neutralizing chemical releasing so that each actuator group has the similar total control input. In Fig. 7.3, the y-axis represents the total pollution all the mesh sensors could detect. Detailed results including $Pollution_{max}$, T_{max} (when the pollution has a peak), and $Pollution_{final}$ are shown in Table 7.3. The case with exactly one static actuator and one pollution source is provided as a baseline for comparison. The robot trajectories are shown in Figs. 7.4–7.6.

7.4.3   Grouping Effect Summary

From the above results, we can see that our CVT algorithm is distributable, scalable and with high performance. All the mobile CVT methods achieve better results than those of the static evenly distributed ones. However, the optimal grouping size for

Table 7.2   Run Time for Simulation and Control Results

Grouping	Actuator numbers	Neutralizing parameter
1*1 Actuator	1	320
2*2 Actuator	4	81
3*3 Actuator	9	36
4*4 Actuator	16	20.25
9*9 Actuator	81	4

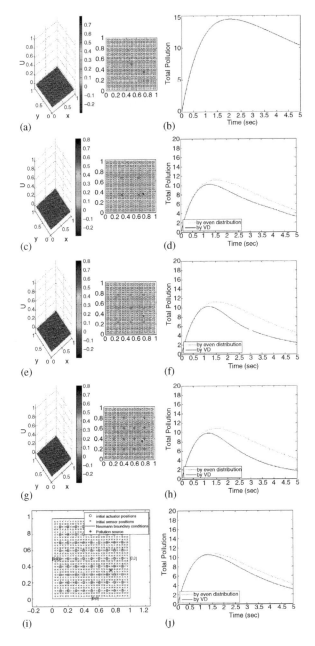

Figure 7.3 Grouping performance for mobile CVT/Static algorithm. (a) Group 1*1 initial layout. (b) Group 1*1 static. (c) Group 2*2 initial layout. (d) Group 2*2 CVT/static. (e) Group 3*3 initial layout. (f) Group 3*3 CVT/static. (g) Group 4*4 initial layout. (h) Group 4*4 CVT/static. (i) Group 9*9 initial layout. (j) Group 9*9 CVT/static.

diffusion control is not always corresponding to the largest size. In other words, under the same total actuation resources, it is *not* definite to tell if the larger number of subgroups corresponds to a better performance. A mathematical model is needed for quantitative analysis of the effect of the grouping size on the efficiency in diffusion control.

Table 7.3 Comparison of Performance for Different Group Size

Grouping	P_{max}	t_{max}	P_{final}	P_{integ}
1*1 Static	100	100	100	100
2*2 (Static)	77.4	74.9	44.4	68.5
2*2 (Mobile)	71.0	59.8	32.0	57.4
3*3 (Static)	77.8	74.3	44.3	68.5
3*3 (Mobile)	71.1	58.4	23.5	50.1
4*4 (Static)	75.3	72.3	41.6	65.9
4*4 (Mobile)	68.2	57.6	16.7	46.3
9*9 (Static)	74.8	73.1	41.2	65.5
9*9 (Mobile)	72.3	63.7	30.2	58.7

Data represented in table are in percentage.

In this section, we extend the application of CVT to the case of large number of mobile actuators for diffusion control. Computational complexity and distributed algorithm are discussed for scalability testing. Through our extensive simulation studies, we demonstrated the effect of the grouping size on the efficiency in diffusion control. Unfortunately, under the same total actuation resources, it is *not* definite to tell if the larger number of subgroups corresponds to a better performance.

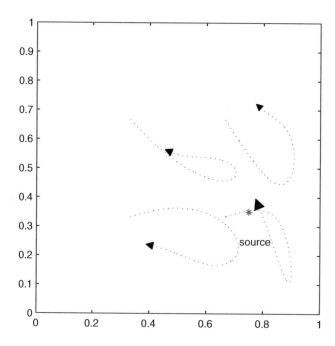

Figure 7.4 Trajectories of 2*2 robots using CVT algorithm.

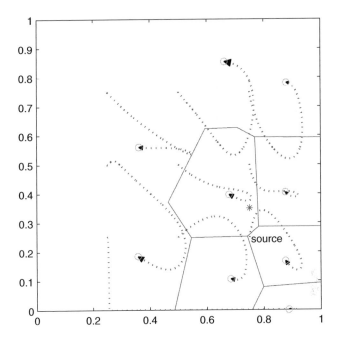

Figure 7.5   Trajectories of 3*3 robots using CVT algorithm and the final Voronoi diagram.

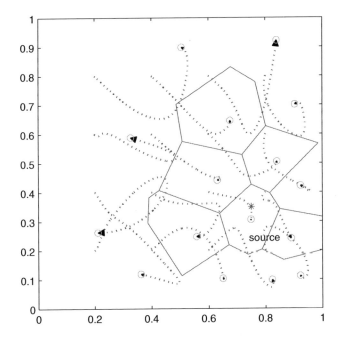

Figure 7.6   Trajectories of 4*4 robots using CVT algorithm and the final Voronoi diagram.

7.5 INFORMATION CONSENSUS IN CVT-BASED DIFFUSION CONTROL

In this section, we introduce information consensus and sharing to the CVT-based diffusion control. The control goal is to drive the actuators to the affected area and counteract the pollutants as quickly as possible.

7.5.1 Basic Consensus Algorithm

First, we review the first-order consensus algorithms [11–13]. Let $p_i \in R^m$ be the information states of the ith robot. For robots with single integrator dynamics given by

$$\dot{p}_i = u_i, \quad i = 1, \ldots, n, \tag{7.11}$$

where $u_i \in R^m$ is the control input, the following first-order consensus algorithm can be applied:

$$u_i = -\sum_{j=1}^{n} g_{ij} k_{ij}(p_i - p_j), \quad i = 1, \ldots, n, \tag{7.12}$$

where g_{ij} represents the set of robots whose information is available to robot i at time t, and k_{ij} is a positive weighting factor.

For the above consensus algorithm, consensus is said to be reached asymptotically among the n vehicles if $p_i(t) \to p_j(t)$, $\forall i \neq j$, as $t \to \infty$ for all $p_i(0)$. A classic rendezvous result is that the rendezvous state can be achieved if the information exchange graph has a spanning tree.

7.5.2 Requirements of Diffusion Control

The pollutant diffusion is both a temporal and a spatial evolution process. CVT method provides a spatial solution to partition the area into small Voronoi diagram and a final state of CVT can be achieved based on different weighted functions. However, the temporal characteristics are also a big challenge for extending CVT to dynamic diffusion control. There are several challenges to incorporate consensus with CVT-based diffusion control.

- *Converging Speed*: To achieve a better control performance, the actuators should converge quickly to the affected area. But all actuators cannot detect the diffusion simultaneously due to the sensing limits. So, the consensus on the affected area needs to be introduced in such a way that the actuators far away from the diffusion source should move faster towards the area with high concentration.

- *Neutralizing Speed*: The final control performance depends highly on how much and where the neutralizing materials are sprayed out. The total amount of the neutralizing material should be minimized given some final constraints on how much to spray totally.
- *Final State*: CVT algorithm (7.6) or (7.8) can guarantee the actuator asymptotically converge to the diffusion source and form a CVT. But this is not enough for diffusion control since a diffusion process evolves with time.

7.5.3 Consensus-Based CVT Algorithm

Based on the above discussions, the new algorithm is proposed for the control of a diffusion process. Consensus algorithm is added on two parts: actuator motion control and actuator spraying/neutralizing control.

Algorithm 7.1

The Consensus-based CVT algorithm is described below:

(1) Initial setting: actuator $p_i \in \{p_1, \cdots, p_n\}$, response time $t = 0$, concentration threshold k_a.

(2) Compute Voronoi region $\bar{V}_i$.

(3) Get the sensor data within the range r_s and compute centroid $\bar{p}_i$ and total pollutant in this region P_{total_i}.

(4) Talk with neighboring actuators. If no diffusion $(\forall i, P_{total_i} < k_a)$, go to 5); else apply corresponding control laws:

    (a) If actuator p_i is out of the affected region ($P_{total_i} < k_a$), make a consensus with neighbors on where is the affected area.

    (b) If actuator p_i is within the affected region ($P_{total_i} > k_a$), make a consensus with neighbors on how fast to spray.

    (c) Else, use CVT control law (7.6) or (7.8).

(5) Stop since no pollution detected.

In the following section, we will explain in detail the two consensus algorithms for motion control and spraying control.

7.5.3.1 Consensus in Actuator Motion Control

In the diffusion process, the actuators sense and react to the diffusion according to the distance from the source. Consensus is introduced here for faster converging speed.

First, the affected area is defined as

$$A_j = \{q \in \Omega | \rho(q) > k_a\} = \{q \in \Omega | |q - d_j| < r_j(t)\}, \qquad (7.13)$$

where d_j is the position of the jth diffusion source, k_a is a positive constant representing the concentration threshold, $r_j(t)$ represents the radius of the affected area. Here we assume there is no wind or other reasons affecting the diffusion process. The consensus to the affected area can be treated as a multileader consensus problem. That is, the actuators out of affected area will follow the actuators already in the affected area. In other words, the diffusion-undetected actuators will follow the diffusion-detected actuators or rendezvous to them until they enter the affected area A_i. The difference with the common "Rendezvous Problem" is that here we need rendezvous to an affected area instead of one point. This can be achieved with disconnected communication topology [13].

$$u_i = - \sum_{j=1}^{n} g_{ij} k_{ij} (p_i - p_j), \quad i = 1, \ldots, n, \qquad (7.14)$$

where $k_{ij} > 0$, $g_{ij} = 0$ and g_{ij} will be set to 1 if information flows from actuator j to i. In our case, it is mostly a leader–follower case. The followers just need to rendezvous to the leaders that are already in the affected area.

Assuming that actuator j is out of the affected area at time t_d, $\mathcal{K}$ should be minimized

$$\frac{\partial \mathcal{K}}{\partial p_j} = 2M_{V_i}(p_i - \bar{p}_i) \simeq 0,$$

$$M_{V_i} \simeq 0. \qquad (7.15)$$

Based on plain CVT actuator motion planning, the actuator j will not react until $|p_i - \bar{p}_i| > \delta$. But the consensus algorithm introduces the information sharing among actuator so that the actuator out of affected area can react early and achieve a faster converging speed.

We set up an emulated scenario to show our idea. Suppose only one actuator (actuator #3) is close to the diffusion source and it detects the diffusion very early Fig. 7.7a. With CVT algorithm, the actuator #3 can drive to the affected area asymptotically. However, other actuators will not react to the diffusion quickly enough since it takes time for the pollutant to enter the area close to other actuators. With consensus algorithm, the actuator #3 can broadcast to the other actuators, or act as the leader of the group and lead all the others into the affected area. In Fig. 7.7b, there are two actuators (#1, #4) that are close to the affected area. So, they will respond to both of the early arrivers and converge to the middle of actuator #1 and #4, which is also the affected area that needs to be controlled or sprayed. With this algorithm, consensus can be reached asymptotically for the n actuators since $p_i - d_j \rightarrow r_j(t)$, as $t \rightarrow \infty$ for all p_i.

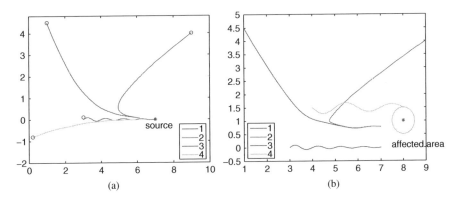

Figure 7.7 Simulation: rendezvous to the affected area. (a) 1 Leader (#3) and 3 Followers (#1, #2, #4). (b) 2 Leaders (#3, #4) and 2 Followers (#1, #2).

7.5.3.2 Consensus in Actuator Neutralizing Control

The plain CVT algorithm introduces a spatial solution to the diffusion control problem [7]. However, the neutralizing control part may not balance. Given a typical pollution/spraying control scenario using the plain CVT algorithm (Fig. 7.8), we can observe from Fig. 7.9 that the actuator #4 sprays more neutralizing chemicals than the total sprayed by the other three, which is not an efficient way when employing more actuators.

In our present study, consensus is introduced to neutralizing control for maximizing the neutralizing ability of every actuator [18]. Consensus is said to be reached for the n actuators if u_{pr_i} is at the same order of magnitude or as close as possible, $\forall i \neq j$, as $t \to \infty$. CVT algorithm (7.6) or (7.8) can guarantee the actuator to converge to a final CVT as $t \to \infty$, but that is a scenario that can not happen in the diffusion evolving scenario. To achieve a better control performance, every actuator should be fully used in the neutralizing control. We wish to use the proposed consensus algorithm to

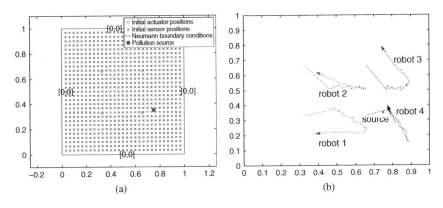

Figure 7.8 Plain CVT diffusion control. (a) Initial positions of actuators. (b) Trajectories of actuators.

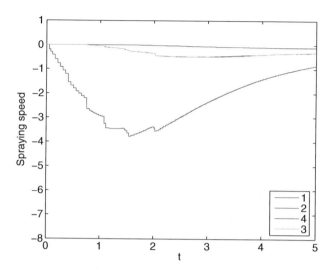

Figure 7.9 Spraying speed comparison for CVT.

avoid the situation that we could not send as many as possible mobile actuators to the most affected area.

To achieve this, the following spraying control input can be applied

$$u_i = -k_p(p_i - \bar{p}_i) - \sum_{j=1}^{N} g_{ij}k_{ij}(p_i - p_j), \tag{7.16}$$

where g_{ij} and k_{ij} have the same definitions as in (7.12). The first part $p_i - \bar{p}_i$ drives the actuator respond to the diffusing and the later part in (7.16) will drive the actuators closer to the actuator that has the highest P_{total_i}.

7.6 SIMULATION RESULTS

Two simulation examples are shown to demonstrate the effectiveness of the new algorithm. The first one has no constraints on how much to spray in total $k_s = \infty$. The second one illustrates how this constraint will affect the final control performance. The implementation details are provided in Appendix D.

Diff-MAS2D [19] is used as the simulation platform for our implementation. The area concerned can be modelled by $\Omega = \{(x, y)|0 \le x \le 1, 0 \le y \le 1\}$. In (7.1) $k = 0.01$ and the boundary condition is given by

$$\frac{\partial u}{\partial n} = 0. \tag{7.17}$$

The stationary pollution source is modeled as a point disturbance f_d to the PDE system (7.1) with its position at (0.8, 0.2) and

$$f_d(t) = 20e^{-t}|_{(x=0.8, y=0.2)}. \tag{7.18}$$

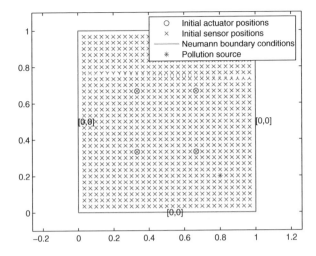

Figure 7.10   Initial layout of diffusion control.

The mesh sensor network is assumed to provide the actuators with measurements on pollutant concentration. There are 29×29 sensors evenly distributed in a square area $(0, 1)^2$ (a unit area) and four mobile actuators/robots that can release the neutralizing chemicals. The pollution source begins to diffuse at $t = 0$ to the area Ω and initially the mobile actuator robots are evenly distributed within the domain Ω (one by one square) at the following specific positions: for 2×2 grouping case, $(0.33, 0.33)$, $(0.33, 0.66)$, $(0.66, 0.33)$, $(0.66, 0.66)$. The actuators and sensors get updates every 0.1 s. The dynamic model of actuator is assumed to be the first order. We will add more simulation results for the second-order model in the final version.

Given the initial layout Fig. 7.10, we need to choose the corresponding control law and communication matrix. Let's consider the vector form of control input:

$$U = L_1 P - L_2 \bar{P}, \tag{7.19}$$

where $U = [u_1^T \cdots u_n^T]$, $P = [p_1^T \cdots p_n^T]$, $\bar{P} = [\bar{p}_1^T \cdots \bar{p}_n^T]$ are all vectors, L_1 is the control matrix determined by communication topology and corresponding control law.

In the beginning, the actuator #3 is relatively close to the diffusion process, and it will detect and react to the diffusing first. Then, it will broadcast this event to all the other three actuators. The communication topology shown in Fig. 7.11 and control matrices L_1 and L_2 are shown below:

$$L_1 = \begin{pmatrix} -1 & 0 & 1 & 0 \\ 0 & -1 & 1 & 0 \\ 0 & 0 & -1 & 0 \\ 0 & 0 & 1 & -1 \end{pmatrix}, \quad L_2 = \begin{pmatrix} 0 & 0 & 0 & 0 \\ 0 & 0 & 0 & 0 \\ 0 & 0 & -1 & 0 \\ 0 & 0 & 0 & 0 \end{pmatrix}.$$

After a certain time, actuator #1 and #4 also enter the affected area. The communication topology and control matrix are then changed. The topology is shown in

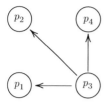

Figure 7.11 p_3 alone broadcasts.

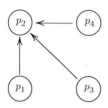

Figure 7.12 p_1, p_3, and p_4 broadcast.

Fig. 7.12 and the new control matrices are

$$L_1 = \begin{pmatrix} -1 & 1 & 1 & 1 \\ 0 & -1 & 0 & 0 \\ 0 & 0 & -1 & 0 \\ 0 & 0 & 0 & -1 \end{pmatrix}, \quad L_2 = \begin{pmatrix} 0 & 0 & 0 & 0 \\ 0 & -1 & 0 & 0 \\ 0 & 0 & -1 & 0 \\ 0 & 0 & 0 & -1 \end{pmatrix}.$$

After all four actuators have entered the affected area, the S_{total_i} are compared and converted to step 4c for consensus on the amounts of neutralizing chemicals. The actuator trajectories are shown in Fig. 7.13.

Figure 7.14 and Table 7.4 show the control performance comparison between plain CVT and consensus-based CVT, which shows a decrease in both the max and final total pollution value. The time the actuators take to arrive at the affected area is compared in Fig. 7.15. It can be observed that consensus-based CVT lead to a

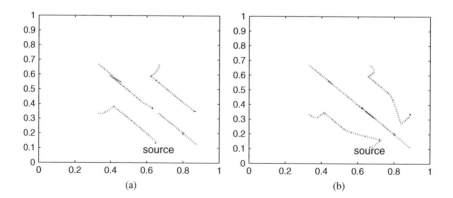

Figure 7.13 Trajectory comparison between the consensus-based CVT and plain CVT. (a) Trajectory of CVT. (b) Trajectory of consensus CVT.

Table 7.4   Comparison of Control Performance

Algorithm	P_{max}	t_{max}	P_{final}
CVT	12.9186	1.7980	1.9330
Consensus CVT	12.7850	1.7420	1.5743
CVT (spray limits)	10.3318	2.3080	4.6901
Consensus (spray limits)	12.7850	1.7420	2.9365

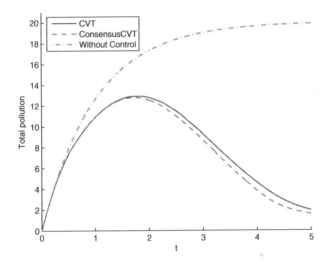

Figure 7.14   Comparison of total pollutants: plain CVT and consensus CVT.

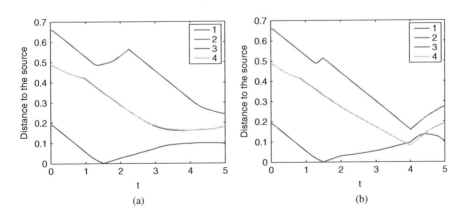

Figure 7.15   Distance to the source. (a) Plain CVT. (b) Consensus CVT.

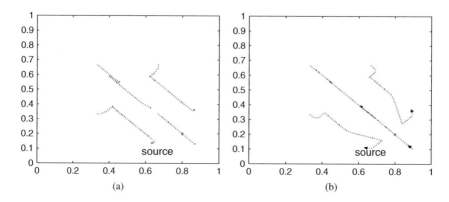

(a)   (b)

Figure 7.16 Actuator trajectories of consensus-based CVT and plain CVT. (a) Plain CVT.
(b) Consensus CVT.

tighter formation during neutralizing. Consensus-based CVT has a better control
performance on the diffusion process over the plain CVT.

When controlling a diffusion process, another important factor is the constraints
on the total neutralizing chemical sprayed (7.3). To make a comparison between
consensus-based CVT and the plain CVT, the total neutralizing amount is reduced to
70% of the preceding case. For consensus-based CVT, a saturation $[-2, 0]$ is added
to guarantee the balance of spraying speed among actuators. The initial layout and
all parameters are the same with the above simulation. The motion trajectories are
shown in Fig. 7.16.

From Fig. 7.17 and Table 7.4, we can observe that although the maximal total
pollutant is smaller, the final pollutant left using plain CVT is 4.6901, which is

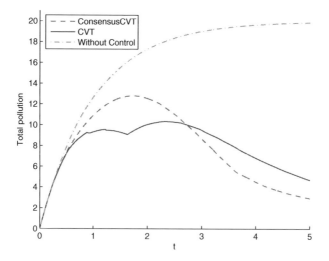

Figure 7.17 Comparison of total pollutants: plain CVT and consensus CVT.

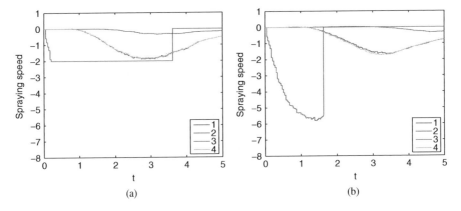

Figure 7.18 Comparison of spraying speeds. (a) Plain CVT. (b) Consensus CVT.

significantly more than that achieved via the consensus-based CVT, which is as low as 2.9365. So, this strategy is not so good because it does not make fully use of the neutralizing ability of all the four actuators.

In summary, the diffusion control problem is quite difficult because it evolves both spatially and temporally, and PDEs are needed for modeling. There remains no good solution based on the authors' best knowledge. With the presented simulation results, the following further discussions are presented in order.

- *Mobile Actuator Control Problem*: One of the difficulties in diffusion control is that both actuator position and neutralizing speed need to be controlled. Especially, the neutralizing control strategy can have a large impact on the final control performance. Different control laws can be designed for various requirements. As shown in Figs. 7.17 and 7.18, CVT algorithm has smaller maximum pollutant values (Table 7.5) but quickly sprays out the total neutralizing chemicals. Consensus CVT outperforms CVT in this aspect because it pays more attention to interactuator communication and tries to maintain a balance of neutralizer amount among actuators.

- *CVT Advantages and Limitations*: CVT algorithm is a non-model-based method to control a diffusion, process and it is easy to implement in a large scale since it needs only the neighbor information. The diffusion source can be moving and can be multiple. However, CVT can only guarantee the slow

Table 7.5 Comparison of Total Neutralizing Material

Algorithm	S_1	S_2	S_3	S_4
CVT	4.25	0.53	9.01	4.18
Consensus CVT	4.47	0.69	8.75	4.42
CVT (spray limits)	3.89	0.31	7.01	4.02
Consensus (spray limits)	4.67	0.70	7.00	4.61

converging to the source, as seen in Fig. 7.15. The final diffusion control performance depends a lot on the initial conditions such as the starting points of actuators. The converging speed and computation burden are also limitations for CVT [5].

- *Communication Topology*: This chapter assumes that the actuator can get the sensor information within a certain distance of effectiveness and a full communication topology among actuators. But the simulation result is only based on some specific communication topologies. Further tests are needed for topology changing or switching while actuator moving and spraying.

- *2D/3D Spatial Problem*: The CVT algorithm is a spatial solution to the diffusion control problem. With the availability of small and powerful robots and sensor network, these kind of spatial problem will sooner or later be solved. Unmanned aerial vehicles will be perfect platforms for this kind of experiments [20].

- *Experimental Validation*: More experiments on real mobile sensor and actuator networks will be interesting to validate the ideas in this chapter.

7.7 CHAPTER SUMMARY

In this chapter, we proposed to incorporate the information sharing and consensus strategy to the CVT-based actuators motion planning for better control of a diffusing process. The new algorithm is tested with a first-order dynamic model and its improvement has been demonstrated, especially under total spraying amount limit. Further simulation results and comparisons can be made in the future using a second-order actuator model.

REFERENCES

1. D. Zarzhitsky, D. F. Spears, and W. M. Spears. Swarms for chemical plume tracing. In *Proceedings of the IEEE Swarm Intelligence Symposium*, pages 249–256, June 2005.

2. M. A. Demetriou. Power management of sensor networks for detection of a moving source in 2-D spatial domains. In *Proceedings of the American Control Conference*, pages 1144–1149, June 2006.

3. Y. Q. Chen, K. L. Moore, and Z. Song. Diffusion boundary determination and zone control via mobile actuator-sensor networks (MAS-net): challenges and opportunities. In *Proceedings of the Intelligent Computing: Theory and Applications II, Part of SPIE's Defense and Security*, 2004.

4. J. Cortés, S. Matínez, T. Karatas, and F. Bullo. Coverage control for mobile sensing networks. *IEEE Transactions on Robotics and Automation*, 20(20):243–255, 2004.

5. Q. Du, V. Faber, and M. Gunzburger. Centroidal Voronoi tessellations: applications and algorithms. *SIAM REVIEW*, 41(4):637–676, 1999.

6. L. Ju, Q. Du, and M. Gunzburger. Probabilistic methods for centroidal voronoi tessellations and their parallel implementations. *Parallel Computation*, 28(10):1477–1500, 2002.

7. Y. Q. Chen, Z. Wang, and J. Liang. Optimal dynamic actuator location in distributed feedback control of a diffusion process. In *Proceedings of the 44th IEEE Conference on Decision and Control*, pages 5662–5667, December 2005.

8. D. Ucinski. *Optimal Measurement Methods for Distributed Parameter System Identification*. CDC Press, Boca Raton, FL, 2004.

9. Y. Q. Chen, Z. Wang, and J. Liang. Automatic dynamic flocking in mobile actuator sensor networks by central Voronoi tessellations. In *Proceedings of the IEEE International Conference on Mechatronics and Automation*, volume 3, pages 1630–1635, August 2005.

10. H. Chao, Y. Q. Chen, and W. Ren. A study of grouping effect on mobile actuator sensor networks for distributed feedback control of diffusion process using central voronoi tessellations. *International Journal of Intelligent Control Systems*, 11(2):185–190, 2006.

11. R. Olfati-Saber and R. M. Murray. Consensus problems in networks of agents with switching topology and time-delays. *IEEE Transactions on Automatic Control*, 49(9):1520–1533, 2004.

12. A. Jadbabaie, J. Lin, and A. S. Morse. Coordination of groups of mobile autonomous agents using nearest neighbor rules. *IEEE Transactions on Automatic Control*, 48(6):988–1001, 2003.

13. W. Ren, H. Chao, W. Bourgeous, N. Sorensen, and Y. Q. Chen. Experimental validation of consensus algorithms for multi-vehicle cooperative control. *IEEE Transactions on Control Systems Technology*, 16(4):745–752, 2008.

14. A. Okabe, B. Boots, and K. Sugihara. *Spatial Tessellations*. 2nd ed. John Wiley, Chicester, UK, 2000.

15. A. Howard, M. J. Mataric, and G. S. Sukhatme. Mobile sensor network deployment using potential fields: a distributed, scalable solution to the area coverage problem. In *Proceedings of the International Symposium on Distributed Autonomous Robotics Systems*, June 2002.

16. N. Heo and P. K. Varshney. Energy-efficient deployment of intelligent mobile sensor networks. *IEEE Transactions on Systems, Man and Cybernetics, Part A*, 35(1):78–92, 2005.

17. C. Bradford Barber, D. P. Dobkin, and H. Huhdanpaa. The quickhull algorithm for convex hulls. *ACM Transaction on Mathematical Software*, 22(4):469–483, 1996.

18. H. Chao, Y. Q. Chen, and W. Ren. Consensus of information in distributed control of a diffusion process using Centroidal Voronoi Tessellations. In *Proceedings of the 46th IEEE Conference on Decision and Control*, pages 1441–1446, December 2007.

19. J. Liang and Y. Q. Chen. Diff-MAS2D (version 0.9) user's manual: a simulation platform for controlling distributed parameter systems (diffusion) with networked movable actuators and sensors (MAS) in 2D domain. Technical Report USU-CSOIS-TR-04-03, CSOIS, Utah State University, 2004.

20. H. Chao, A. M. Jensen, Y. Han, Y. Q. Chen, and M. McKee. *Advances in Geoscience and Remote Sensing*, chapter AggieAir: Towards low-cost cooperative multispectral remote sensing using small unmanned aircraft systems. IN-TECH, Vukovar, Croatia, 2009.

Chapter 8

Conclusions and Future Research Suggestions

8.1 CONCLUSIONS

This monograph provides new approaches on using unmanned vehicles for remote sensing and distributed control applications. The goal is to bridge the gap between the estimation and control theories of unmanned systems and real environmental or agricultural applications. The AggieAir UAS platform is first introduced in Chapter 2 with a detailed small UAV system review, the subsystem design, the IMU interface design, and the whole system integration tests. The big challenge here is to provide a low-cost yet robust solution since most agricultural and environmental scenarios can not afford expensive UAVs, which are usually more than $10,000. Our AggieAir2 platform has shown great performance for remote sensing applications with the lowest possible price ($< \$5000$ for all the hardware), to the author's best knowledge. With the robust platform, more algorithms have been tested to get a better estimation and control performance. In Chapter 3, the state estimation algorithm is focused because of its importance for the flight control and the later georeferencing. Several typical state estimation filters are tested on our own flight data, including different kinds of extended Kalman filters. Chapter 4 is about an advanced lateral flight controller design. The fractional calculus-based techniques are used on the roll channel of the flight control system. A designed PI^α controller show better tracking performance than the designed PID controller both in simulations and real flights. Then, the single and multiple UAV-based remote sensing problems are focused in Chapters 5 and 6, respectively. The path planning subproblem and the georeferencing problem are solved separately. Chapter 7 is devoted to the diffusion control problem using mobile sensor and actuator networks.

Remote Sensing and Actuation Using Unmanned Vehicles, First Edition. Haiyang Chao and YangQuan Chen.
© 2012 by The Institute of Electrical and Electronics Engineers, Inc.
Published 2012 by John Wiley & Sons, Inc.

8.2 FUTURE RESEARCH SUGGESTIONS

Many challenges remain when we apply unmanned systems for civilian applications. Our successful AggieAir UAV platform uses the fixed-wing airframe. However, there are scenarios such as chemical leaking in a factory or urban areas, which requires a more flexible aerial platform; for example, a helicopter-like UAV, or vertically take-off and landing (VTOL) UAV.

8.2.1 VTOL UAS Design for Civilian Applications

VTOL UAVs can fly at extremely low heights with a low ground speed or even stay at a certain position in hovering mode. They could take off and land more easily, independent of runways or complex terrains, which make them especially useful for indoor and surface applications. VTOL UAVs have the following advantages over traditional fixed-wing UAVs.

- Easy for launching and landing even among buildings, on boats, or on moving trucks.
- Much better vertical maneuverability.
- Hovering capability, which is good for image capturing.
- Capable of both very slow and fast flying speed.
- Safer to manipulate and more portable (no wings).

With all the above specialties, VTOL UAVs have many potential civilian applications especially in complex urban areas, which may not be suitable for fixed-wing UAVs since a closer aerial view is required. There are many example tasks.

- *Inspection of Outdoor or Indoor Structures*: Aerial images collected by VTOL UAVs from multiple views can be used for inspection of dams, bridges, tunnels, or power lines instead of sending human beings there.
- *Transportation Applications*: VTOL UAVs can provide big advantages to routine works of the department of transportation. Sample applications include road health monitoring, traffic surveillance, construction monitoring, and so on. High-resolution images taken from VTOL UAVs can even help reconstruct the traffic accident scene.
- *Accurate 3D Measurements*: The omnidirectional mobility of the VTOL UAV increases its flexibility for measurements in a 3D space. For instance, VTOL UAVs can be used to measure the vertical wind speed more accurately than fixed-wing UAVs since they can hover in the air.
- *Thermal Audit*: VTOL UAVs with thermal imagers can be used to determine where the heat leaking of a building is for energy saving purposes.

- *High-Resolution and High-Quality Photos and Videos in the Air*: VTOL UAVs can be easily carried around, launched, and sent to any altitude around the building or the scene for an aerial view.
- *Education and Entertainment*: VTOL UAVs can be demonstrated or used even inside buildings because they do not require much space for take-off and landing.

8.2.2 Monitoring and Control of Fast-Evolving Processes

Fast-evolving processes often need faster information collection and control decision, which is especially suitable for unmanned systems. Both the fog monitoring and the diffusion control problem are only tested in the DiffMAS2D software simulation platform so far. It will be interesting if we can use UAVs to monitor a real diffusion process. The development of algorithms regarding emergence responses will be very useful for dangerous missions like forest fires monitoring or chemical leaks predictions.

In addition, UAVs can form groups for these missions to provide even quicker responses. UAVs flying in meshes or cubes provide more real-time information compared with a single UAV, shown in Fig. 8.1 [1]. They can even coordinate with each other on variables like the vehicle distance or formation shape based upon requirements from different missions.

8.2.3 Other Future Research Suggestions

There are several other future research suggestions regarding the UAV project including

- *Risk Analysis*: The robustness analysis of small UAVs is still a new topic. The probability of failure for small UAV manipulations needs to be modeled and analyzed for flight safety requirements.

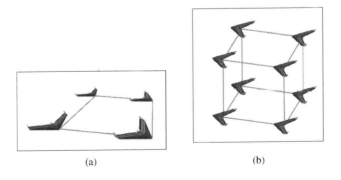

(a)   (b)

Figure 8.1   UAVs in formation [1]. (a) Flying mesh [1]. (b) Flying cube [1].

- *UAV as the Actuator*: Most current UAV applications are for surveillance and monitoring purposes. However, UAV can also work as an actuator. For instance, a VTOL UAV can be used for soil sampling missions with a manipulator installed.

- *Vision-Aided Navigation*: Most of the current UAV platforms rely on GPS/INS as the navigation solution. However, there are also many scenarios where GPS signals may be degraded or even unavailable. Such cases include indoor navigation, canyon or forests navigation, or battle field flights. Given the fact that birds and insects use vision to navigate for millions of years, similar vision-aided navigation techniques will become available to UAVs very soon.

- *New UAV Design to Use Fuel Cell Batteries*: The current power system for AggieAir UAS uses Li-Polymer batteries, with the power density of about 185 Wh/kg. However, the power density of the new fuel cell batteries could be as high as 450 Wh/kg [2]. The usage of fuel cell battery can greatly increase the UAV flight time.

- *Power Consumption Minimization*: The current AggieAir system is stable but not optimal on the power consumption. Both the flight control strategies and the airframe designing can be optimized to get more flight time.

REFERENCES

1. A. M. Jensen, D. Morgan, S. Clemens, Y. Q. Chen, and T. Hardy. Using multiple open-source low-cost unmanned aerial vehicles (UAV) for 3D photogrammetry and distributed wind measurement. In *Proceedings of the ASME Design Engineering Technical Conference Computers and Information in Engineering*, number MESA-87586, September 2009.

2. Horizon Fuel Cell Techonologies. Fuel cell battery for UAVs, 2009. http://www.hes.sg/.

Appendix

A.1 LIST OF DOCUMENTS FOR CSOIS FLIGHT TEST PROTOCOL

A series of flight test documents are included in this appendix to show how to perform routine flight tests in a systematic way. One of the contributions of this monograph is to establish the flight test protocol to minimize the happening of accidents. The flight test documents are summarized based on the extensive flight test experiences from CSOIS OSAM-UAV team.

A.1.1 Sample CSOIS-OSAM Flight Test Request Form

(1) Flight Test Time: 10/05/2009 8:00 - 12:00 AM

(2) Field Test Crew: Haiyang Chao, Ying Luo, Hu Sheng

(3) Flight Test Missions (in priorities): Fly ChangE with Aggie-controller 0.2
  - Square: roll_desire $= -10 - +10$, pitch_desire $= 0$, throttle_desire $= 0.7$. Square_time $= 5*2$ s.
  - Square: roll_desire $= -20 - +20$, pitch_desire $= 0$, throttle_desire $= 0.7$. Square_time $= 5*2$ s.
  - PRBS: pitch_desire $= 0$, throttle_desire $= 0.7$. Roll_design_Length $= 15$.
  - Outer_loop test with P controller.
  - Outer_loop test with PI controller.

(4) Field Operators Required:
  - Flight Test Preparation and Checklist: Haiyang Chao, Ying Luo, Hu Sheng
  - Car Rental: UWRL truck driven by A.M. Jensen
  - Safety Pilot: Haiyang Chao
  - GCS Operator: Hu Sheng (Ying Luo to learn)
  - Camera Record: Ying Luo

Remote Sensing and Actuation Using Unmanned Vehicles, First Edition. Haiyang Chao and YangQuan Chen.
© 2012 by The Institute of Electrical and Electronics Engineers, Inc.
Published 2012 by John Wiley & Sons, Inc.

(5) Airplanes needed: ChangE 60 in.

(6) Lab Preparation Needed:
- Special test for this mission: Check Aggie-controller 0_2 on the ground
- In-lab airframes inspection
- What-to-bring checklist

A.1.2 Sample CSOIS-OSAM 48 in. UAV (IR) In-lab Inspection Form

UAV Inspector			
UAV	Name	Pheonix 48 in.	Dimon 48 in.
ID	Take-off weight (kg)		
	RC channel		
System	Surface smooth		
Check	Winglet stable		
Balance	Hang-up		
Check	Manual		
Actuators	Default trim (1/30 in.)		
Check	Elevon moving range		
	Propeller no damage		
	Throttle moves correctly		
	Elevon moves correctly		
	Nothing loose		
Sensors	IR sensor check		
Check	GPS quick lock		
	Correct orientation (PFD)		
	Auto1 check		
Communication	RC range (step)		
Check	Modem commit		

A.1.3 Sample Preflight Checklist

UAV Tiger Preflight Checklist for AUVSI SUAS Competition 2009

(1) Motor mount screws tight

(2) Prop bolt tight

(3) Nothing loose

(4) Elevons move correctly

(5) Throttle turns on

(6) Correct IMU data (orientation)

(7) GPS lock valid

(8) WiFi link OK

(9) Camera lens cover clear

(10) Camera setup OK (fire up altitude and camera setting)

(11) Camera status (1 means initialized)

(12) Gumstix SD card clear

(13) Walk 25 big steps into the wind after the bungee is fully extended

(14) Turn on the throttle switch right before the launch. Raise the UAV as high as you can and let it go.

A.2 IMU/GPS SERIAL COMMUNICATION PROTOCOLS

Most stand-alone IMUs or GPS units communicate with other devices through serial ports, RS232, UART, or RS485. The example communication protocols are collected in the appendix including u-blox GPS, several CTOS IMUs for the convenience of other unmanned system developers.

A.2.1 u-blox GPS Serial Protocol

U-blox GPS modules have many series such as LEA-4p, LEA-5h, and so on. Most modules support GPS update rate at 1 Hz. Some modules may support update rate at 4 Hz. The UBX GPS messages required for Paparazzi autopilot include [1]:

- UBX_NAV_POSLLH_ID, for latitude, longitude, and altitude
- UBX_NAV_VELNED_ID, for v_n, v_e, v_d, Heading, GSpeed, and ITOW
- UBX_NAV_SOL_ID, for Pacc, Sacc, Pdop, and NumSV
- UBX_NAV_SVINFO_ID, for the satellite information.

The UBX communication protocol with GPS module is defined as follows:

Bytes	1	2	3	4	5	6	6 ~ 5 + LEN	6 + LEN	7 + LEN
	B5	62	CLASS	ID	LEN1	LEN2	PAYLOAD	CK1	CK2

A.2.2 Crossbow MNAV IMU Serial Protocol

Crossbow MNAV IMU has a complete set of sensors for unmanned system applications, shown in Fig. A.1 [2]. The following packets need to be sent to the MNAV to start the scaled mode with full sensor packet [2,3].

(1) Set up baud rate as 57,600 bps (0x55, 55, 57, 46, 01, 00, 02, 00, 03, 00, A3),

(2) Set sampling frequency as 50 Hz (0x55, 55, 53, 46, 01, 00, 01, 00, 02, 00, 9D),

(3) Set up scaled mode (0x55, 55, 53, 46, 01, 00, 03, 00, 53, 00, F0).

Figure A.1 Crossbow
MNAV IMU [2].

The serial communication protocol with Crossbow MNAV IMU is defined as follows:

Bytes	1	2	3	4 ∼ 3 + LEN	4 + LEN	5 + LEN
	55	55	ID	PAYLOAD	CK1	CK2

The complete data packets have the ID of "S" and "N," representing the IMU packet without and with GPS, respectively.

A.2.3 Microstrain GX2 IMU Serial Protocol

Microstrain GX2 IMU can provide orientation information at up to 250 Hz with typical accuracy of 2 degrees under dynamic test conditions [4]. It supports hardware connections through RS232, RS485, RS422, USB, or wireless [4]. The serial baud rate is 115,200 bps, which permits packets up to 115 byte in 100 Hz or packets up to 230 byte in 50 Hz. A set of communication commands can be sent to GX2 IMU to switch between polled mode and continuous mode. The default polled mode means the IMU will transmit a data record only when a command byte is issued from the host computer. The continuous mode means the data are output continuously with no host intervention. The continuous mode and the CC packet are chosen by the authors for GhostGX2 including the acceleration, angular rate, magnetometer vectors, and the rotation matrix (totally 79 bytes including one header byte and two checksum bytes). The CC continuous packet is triggered by sending one C4 command packet (including four bytes: C4, C1, 29, CC) when GhostGX2 starts. The IMU will respond with a 8-byte packet for acknowledgement and then enter the continuous mode to output 79-byte packet at preset rate. The example 79-byte serial packet definition from Microstrain GX2 IMU to the host computer is defined as follows:

Bytes	1	2~ 73	74	75	76	77	78	79
	SYNC	PAYLOAD	T1	T2	T3	T4	CK1	CK2

Several functions are defined to finish the serial communication and parsing tasks include the following:

- Function gx2_read8bytes(), read the ACK packet from IMU in response to the C4 command.
- Function gx2_init(), open the serial port, and send out the initialization packet to set the GX2 IMU in continuous mode.
- Function gx2_close(), close the serial port.
- Function parse_gx2(), read data bytes from serial and assemble them into a CC packet based on the definition.
- Function parse_gx2_cc_msg(), convert the received binary packet to variables in engineering units.

The source codes for the above implementation are provided in the following. The definition of function FloatFromBytes can be found in the product CD by Microstain.

```c
/*Simplified main.c for communication with GX2 IMU */

int main(int argc, char **argv) {
        gx2_init();

        while(1){
            while(1!=read(sPort1,input_buffer,1));
            parse_gx2(input_buffer[0]);

            if (gx2_msg_received) {
                Count_gx2++;
                parse_gx2_cc_msg();
                gx2_msg_received = false;
            }
        }
}

/*  File description: com_gx2.c is for serial communication
between gumstix and Microstrain gx2 IMU.
** Author: Haiyang Chao & Cal Coopmans 20080924 */
#include "globaldef.h"
#include "serial.h"
#include "ser_data.h"
#include "com_gx2.h"
#include <unistd.h>
#include <math.h>

int sPort1; /*serial port to microstrain g2 imu*/
volatile bool gx2_msg_received;
struct imu2pprz ugearpacket_imu;
struct gx2_imu_log *imu_data_ptr;
```

```
gx2_cc_data.M11 = FloatFromBytes(gx2_msg_buf+37);
gx2_cc_data.M12 = FloatFromBytes(gx2_msg_buf+41);
gx2_cc_data.M13 = FloatFromBytes(gx2_msg_buf+45);
gx2_cc_data.M21 = FloatFromBytes(gx2_msg_buf+49);
gx2_cc_data.M22 = FloatFromBytes(gx2_msg_buf+53);
gx2_cc_data.M23 = FloatFromBytes(gx2_msg_buf+57);
gx2_cc_data.M31 = FloatFromBytes(gx2_msg_buf+61);
gx2_cc_data.M32 = FloatFromBytes(gx2_msg_buf+65);
gx2_cc_data.M33 = FloatFromBytes(gx2_msg_buf+69);
gx2_cc_data.lTimer = ULongFromBytes(gx2_msg_buf+73);

gx2_cc_data.cChecksum[0] = gx2_msg_buf[77];
gx2_cc_data.cChecksum[1] = gx2_msg_buf[78];
gx2_cc_data.CPUTIME = (float)get_Time();
pitch = asin((-1)*gx2_cc_data.M13/R2D)/PI*180;
roll = atan2(gx2_cc_data.M23,gx2_cc_data.M33)/PI*180;
yaw = atan2(gx2_cc_data.M12,gx2_cc_data.M11)/PI*180;
ugearpacket_imu.the = (int16_t)(asin((-1)*gx2_cc_data.M13/R2D)*10000);
ugearpacket_imu.phi = (int16_t)(atan2(gx2_cc_data.M23,gx2_cc_data.M33)*10000);
ugearpacket_imu.psi = (int16_t)(atan2(gx2_cc_data.M12,gx2_cc_data.M11)*10000);

/*add gps_log to log all the data in integer, */
gx2_log.the = ugearpacket_imu.the;
gx2_log.phi = ugearpacket_imu.phi;
gx2_log.psi = ugearpacket_imu.psi;
gx2_log.AccelX = (int32_t)(gx2_cc_data.fAccelX*10000);
gx2_log.AccelY = (int32_t)(gx2_cc_data.fAccelY*10000);
gx2_log.AccelZ = (int32_t)(gx2_cc_data.fAccelZ*10000);
gx2_log.RateX = (int32_t)(gx2_cc_data.fRateX*10000);
gx2_log.RateY = (int32_t)(gx2_cc_data.fRateY*10000);
gx2_log.RateZ = (int32_t)(gx2_cc_data.fRateZ*10000);
gx2_log.MagX = (int32_t)(gx2_cc_data.fMagX*10000);
gx2_log.MagY = (int32_t)(gx2_cc_data.fMagY*10000);
gx2_log.MagZ = (int32_t)(gx2_cc_data.fMagZ*10000);
gx2_log.CPUTIME = (uint32_t)(get_Time()*10000);
imu_data_ptr = &gx2_log; /*add on 20091112 for logging in separate thread*/

/*  imu logging */
if ( ((i_display++) % 2) == 0)
    fwrite(&gx2_log, sizeof(struct gx2_imu_log),1,fimu);
if ((i_display % 50) == 0){
    if ( fflush(fimu) != 0 )
        printf("imu data log flush error!\n");
    if ( fsync(fileno(fimu)) != 0)
        printf("imu data log sync error!\n");
    i_display = 0;
    }
}
```

A.2.4 Xsens Mti-g IMU Serial Protocol

Xsens Mti-g IMU is a GPS-aided IMU with GPS, static pressure, and inertial sensors [5], shown in Fig. A.2. It uses an extended Kalman filter to provide orientation

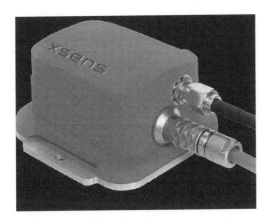

Figure A.2   Xsens Mti-g IMU [5].

estimation with 1 degree RMS accuracy under dynamic cases [5]. The serial link of Xsens Mti-g IMU can be set either in config state or in measurement state. Config state is for configuration settings while measurement state is for continuous data output. GoToConfig or GoToMeasurement commands can be used to switch between these two modes. The Xsens MT serial packet is defined as follows:

Bytes	1	2	3	4	4~ 3 + LEN	4 + LEN
	PRE	BID	MID	LEN	PAYLOAD	CHECKSUM

PRE stands for preamble, or the packet header; BID stands for bus identifier or address; MID stands for message identified; LEN stands for the number of bytes in PAYLOAD field. For example, GoToConfig message includes five bytes: 0xFA, FF, 30, 00, CK. GoMeasurement message also includes five bytes: 0xFA, FF, 10, 00, CK. The whole initialization procedure can be divided into several steps:

- Host Computer → Mti-g: GoToConfig (0xFA, FF, 30, 00, CK).
- Wait for acknowledgement from Mti-g (0xFA, FF, 31).
- Host Computer → Mti-g: SetOutputMode (0xFA, FF, D0, 02, 08, 36, CK). The IMU is set to output calibrated orientation, position, velocity, and status data.
- Wait for acknowledgement from Mti-g (0xFA, FF, D1).
- Host Computer → Mti-g: SetOutputSettings (0xFA, FF, 04, 80, 00, 0C, 05, CK). The IMU is set to output Euler angles and output in float format.
- Wait for acknowledgment from Mti-g (0xFA, FF, D3).
- Host Computer → Mti-g: GoToMeasurement (0xFA, FF, 10, 00, CK).
- Wait for acknowledgment from Mti-g (0xFA, FF, 11).

The source codes for the initialization of Xsens Mti-g IMU are also provided in below.

```
void xsens_init( void ) {

    int         nbytes = 0;
    uint8_t     GoToConfig[5]          ={0xFA,0xFF,0x30,0x00,0x00};
    uint8_t     SetOutputMode[7]       ={0xFA,0xFF,0xD0,0x02,0x08,0x36,0x00};
    uint8_t     SetOutputSettings[9]   ={0xFA,0xFF,0xD2,0x04,0x80,0x00,0x0C,0x05,0x00};
    uint8_t     GoToMeasurment[5]      ={0xFA,0xFF,0x10,0x00,0x00};
    char        xsens_dev[64]          ="/dev/ttyS1";
    uint8_t     cksum[2]               ={0x00,0x00};
    uint8_t     GoToConfig_full[5]     ={0xFA,0xFF,0x30,0x00,0xD1};
    uint8_t     ReqDID_full[5]         ={0xFA,0xFF,0x00,0x00,0x01};
    bool        ack_received           = false;
    int         ack_count              = 0;
    uint8_t     ack_buf[4]             ={0x00,0x00,0x00,0x00};

    ins_msg_received == false;
    msg2pprz_ready = false;
    xsens_status = UNINIT;

    sPort = open_serial( xsens_dev, BAUDRATE_115200, 0 );
    #ifdef LOG_ON
        printf("sPort: %d, opened serial port at 115200.\n",sPort);
    #endif

    nbytes = console_write(GoToConfig_full,5);
    printf("\n GoToConfig: \n");
    while( ack_received != true ){
        if ( (ack_buf[2] == 0xFA) && (ack_buf[1] == 0xFF) && (ack_buf[0] == 0x31) ){
            ack_received = true;
        }
        while ( 1 != read(sPort,&xsens_msg_buf[0],1) );

        for (ack_count = 0; ack_count < 3; ack_count ++){
            ack_buf[3 - ack_count] = ack_buf[2 - ack_count];
        }
        ack_buf[0] = xsens_msg_buf[0];
    }
    ack_received = false;
    for (nbytes = 0; nbytes < ack_buf[0]+1; nbytes++){
        while ( 1 != read(sPort,&xsens_msg_buf[0],1) );
    }

    SetOutputMode[6] = xsens_cksum(SetOutputMode,6);
    nbytes = console_write(SetOutputMode,7);
    printf("\n SetOutputMode: \n");
    while( ack_received != true ){
        if ( (ack_buf[2] == 0xFA) && (ack_buf[1] == 0xFF) && (ack_buf[0] == 0xD1) ){
            ack_received = true;
        }
```

```
    while ( 1 != read(sPort,&xsens_msg_buf[0],1) );

    for (ack_count = 0; ack_count < 3; ack_count ++){
        ack_buf[3 - ack_count] = ack_buf[2 - ack_count];
    }
    ack_buf[0] = xsens_msg_buf[0];
}
ack_received = false;
for (nbytes = 0; nbytes < ack_buf[0]+1; nbytes++){
    while ( 1 != read(sPort,&xsens_msg_buf[0],1) );
    //printf("%X, ", xsens_msg_buf[0]);
}

SetOutputSettings[8] = xsens_cksum(SetOutputSettings,8);
nbytes = console_write(SetOutputSettings,9);
printf("\n SetOutputSettings: \n");
while( ack_received == false ){
    if ( (ack_buf[2] == 0xFA) && (ack_buf[1] == 0xFF) && (ack_buf[0] == 0xD3) ){
        ack_received = true;
    }
    while ( 1 != read(sPort,&xsens_msg_buf[0],1) );

    for (ack_count = 0; ack_count < 3; ack_count ++){
        ack_buf[3 - ack_count] = ack_buf[2 - ack_count];
    }
    ack_buf[0] = xsens_msg_buf[0];
}
ack_received = false;
for (nbytes = 0; nbytes < ack_buf[0]+1; nbytes++){
    while ( 1 != read(sPort,&xsens_msg_buf[0],1) );
    //printf("%X, ", xsens_msg_buf[0]);
}
GoToMeasurment[4] = xsens_cksum(GoToMeasurment,4);
nbytes = console_write(GoToMeasurment,5);
printf("\n GoToMeasurment: \n");
while( ack_received == false ){
    if ( (ack_buf[2] == 0xFA) && (ack_buf[1] == 0xFF) && (ack_buf[0] == 0x11) ){
        ack_received = true;
    }
    while ( 1 != read(sPort,&xsens_msg_buf[0],1) );

    for (ack_count = 0; ack_count < 3; ack_count ++){
        ack_buf[3 - ack_count] = ack_buf[2 - ack_count];
    }
    ack_buf[0] = xsens_msg_buf[0];
}
ack_received = false;
for (nbytes = 0; nbytes < ack_buf[0]+1; nbytes++){
    while ( 1 != read(sPort,&xsens_msg_buf[0],1) );
    //printf("%X, ", xsens_msg_buf[0]);
}
printf("Finished all the settings! \n");
}
```

A.3 PAPARAZZI AUTOPILOT SOFTWARE ARCHITECTURE: A MODIFICATION GUIDE

This guide is to explain the structure of the Paparazzi autopilot software for further modification purposes based on the author's working experiences. The purpose of this guide is to facilitate the readers who are interested in adapting the Paparazzi autopilot for their own developments. The major function of Paparazzi autopilot software is to guide the UAV for the autonomous flight [6]. The autopilot code is written with the C language since it runs on embedded systems such as LPC2148. For the airframe- and mission-specific configurations, the XML files are used including flight plan setting, telemetry selection, controller tuning, and so on. The compiler will convert all these XML files into C files for the code uploading.

A.3.1 Autopilot Software Structure

In order to achieve the autonomous navigation task, the autopilot needs several sub-functions such as collecting data from sensors, actuating control surfaces based on the flight plan, and communicating with the ground for health monitoring and emergency responses. The whole Paparazzi autopilot software structure is shown in Fig. A.3. For the AggieAir2 UAV platform, the sensors include the GPS, IMU, and the voltmeter while the actuators include the throttle motor and the elevons. The communication devices include one 900-MHz serial modem and one RC receiver. The remote control using the RC transmitter is also called the safety link, or fly-by-wire (FBW). The data link between the UAV and the ground is connected through a pair of modems. The data link includes the telemetry for health monitoring (one-way communication) and the DL setting for real-time tunings (two-way communication). The autopilot software is to achieve the sensing, actuation, and communication tasks using all the

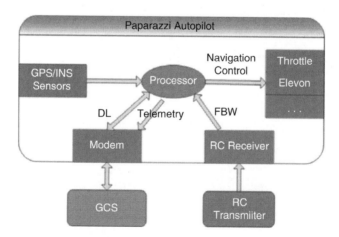

Figure A.3 Paparazzi autopilot software structure.

devices described above. The codes running on the central processor is written in C and the specific parameters for control and navigation purposes are implemented in XML files [6].

A.3.2 Airborne C Files

The key parts of the Paparazzi autopilot code are implemented in C. The airborne files are located in the /*paparazzi*3/*sw*/*airborne* directory. Because Paparazzi UAV is an open-source project, many macros and conditional inclusions (#*ifdef*, #*endif*) are used for different users. A brief description of the airborne files is given in the following:

- *main.c*: the main function of the whole autopilot code
- *main_ap.c*: the subfunctions implemented with either time interrupts or event interrupts
- *main_fbw.c*: the fly-by-wire implementation for the RC control
- *ap_downlink.h*: the communication subfunction definitions
- *gps.c* and *gps_ubx.c*: the GPS protocol
- *osam_ugear.c* and *osam_ugear.h*: the OSAM-Paparazzi interface for getting GPS/INS data from the Gumstix
- *infrared.c*: to read the Infrared sensor outputs through ADC
- *fw_h_ctl.c* and *fw_v_ctl.c*: for the low-level control.

Paparazzi software uses a "task event" structure [6]. For example, the "main.c" is implemented as follows.

Paparazzi airborne code: main function.

```
while(1){
        if (sys_time_periodic()) {
                Fbw(periodic_task);
                Ap(periodic_task);
                Fbw(event_task);
                Ap(event_task);
        }
        ...
}
```

The Paparazzi UAV project provides several demo examples in the airborne directory for a better understanding of the airborne code writing using LPC processors. The demo file examples include [6]:

- *main_demo1.c*: to show how the code works with no specific functions
- *main_demo2.c*: to show how to make the LED toggle

- *main_demo3.c*: to make the LED toggle and send data to UART0
- *main_demo4.c*: to downlink formatted messages to the modem UART
- *main_demo5.c*: to show how to modify the setting files through a user defined GUI
- *main_demo6.c*: for the USB-serial communication.

All the XML files are first converted to C header files. Then the airborne files are linked and compiled into an elf file. The elf file is finally uploaded to the autopilot.

A.3.3   OSAM-Paparazzi Interface Implementation

The functions of the OSAM-Paparazzi interface are explained in detail in Chapter 2. This manual focuses on the implementation part. Several C files are modified or added to the airborne directory.

- *main_ap.c*: modified to add the new event of receiving packets from Gumstix
- *ap_downlink.h*: modified to add new telemetry messages for debugging purposes
- *infrared.c*: modified to stop the angle updates from infrared sensors
- *osam_imu_ugear.c* and *osam_imu_ugear.h*: created for the OSAM-Paparazzi communication protocol.

The event of receiving messages from the Gumstix is added to replace the default event to receive the GPS messages.

Event modifications in *main_ap.c*

```
#ifdef OSAMUGEAR
    if (UgearBuffer()) {
        ReadUgearBuffer();
    }
    if (ugear_msg_received){
        parse_ugear_msg();
        ugear_msg_received = false;
        if (gps_pos_available){
            UseGpsPosUgear(estimator_update_state_gps);
            gps_pos_available = FALSE;
            . . .
        }
    }
#endif /* OSAMUGEAR*/
```

The airframe configuration XML file needs also to be modified since the OSAM-Paparazzi interface is used to replace the default GPS for the UART1. The following lines need to be added.

```
ap.CFLAGS += -DOSAMUGEAR -DUGEAR_LED=2 -DUSE_UART1
-DOSAMUGEAR_LINK=Uart1 -DUART1_BAUD=B115200

ap.srcs += osam_ugear.c latlong.c
```

A.3.4 Configuration XML Files

Paparazzi UAV project uses several XML files for the user-specific configuration settings including the flight plan, airframe parameters, and communication setting parameters [6]. Different users could have different navigation plans, various airframes or communication devices. The major configuration files include the following:

- The airframe configuration file, which contains all the airframe parameters such as controller gains, elevon mix setting, max servo limits, and most importantly, the airborne files for uploading.
- The flight plan, which is used to set the default flight plan and exceptional behaviors using blocks.
- The data link setting, which is used to specify the DLsetting message types (DLsetting is used for online tuning).
- The telemetry configuration file, which is used to set the telemetry messages to be sent to the GCS in real time.
- The radio configuration file, which contains the RC transmitter setting.

All the configuration files are stored in the $/paparazzi3/conf$ directory or subdirectories under it by default.

An example airframe configuration file for a flying-wing UAV is provided in the following part. It comprises two parts: the first part is on the control parameters and the second part is on the airborne files to be included. Some of the control parameters can be chosen and tested on the ground. For example, AILEVON_RIGHT describes the PWM signal generated to drive RC servo. Those variables, which are directly related to flight performance, require in-flight fine tuning. Such variables include the following:

- AUTO1 MAX_ROLL and MAX_PITCH, represent the maximal values we can set in the inner-loop attitude tracking.
- HORIZONTAL CONTROL variables, represent the PID gains used for horizontal control of airplanes.
- VERTICAL CONTROL variables, represent the PID gains used for vertical control of airplanes.
- AGGRESSIVE variables, represent the PID gains used in aggressive mode, especially during take-off.

The airframe configuration file of AggieAir 72 in. Tiger is provided in below. Tiger communicates with a MaxStream 9XTend modem on UART1 with the baud rate at 9600 bps and communicates with Gumstix on UART0 with the baud rate at 115,200 bps. The extra file (osam_imu_ugear.c) is included for specific serial protocol.

```xml
<!-- This file is modified by Haiyang Chao for Communication with
Gumstix + Microstrain GX2 + Ublox GPS Lea5h-->

<airframe name="OSAM72TIGER">

<!-- commands section -->
  <servos>
    <servo name="AILEVON_RIGHT" no="3" min="1900" neutral="1450" max="1000"/>
    <servo name="AILEVON_LEFT"  no="4" min="1050" neutral="1450" max="1900"/>
    <servo name="THROTTLE"      no="7" min="1003" neutral="1003" max="1875"/>
  </servos>

  <commands>
    <axis name="THROTTLE" failsafe_value="0"/>
    <axis name="ROLL"     failsafe_value="0"/>
    <axis name="PITCH"    failsafe_value="0"/>
  </commands>

  <rc_commands>
    <set command="THROTTLE" value="@THROTTLE"/>
    <set command="ROLL"     value="@ROLL"/>
    <set command="PITCH"    value="@PITCH"/>
  </rc_commands>

  <section name="MIXER">
    <define name="AILEVON_AILERON_RATE" value="0.8"/>
    <define name="AILEVON_ELEVATOR_RATE" value="0.7"/>
  </section>

  <command_laws>
    <let var="aileron"         value="-@ROLL  * AILEVON_AILERON_RATE"/>
    <let var="elevator"        value="-@PITCH * AILEVON_ELEVATOR_RATE"/>
    <set servo="THROTTLE"      value="@THROTTLE"/>
    <set servo="AILEVON_LEFT"  value="$elevator + $aileron"/>
    <set servo="AILEVON_RIGHT" value="$elevator - $aileron"/>
  </command_laws>

  <section name="AUTO1" prefix="AUTO1_">
    <define name="MAX_ROLL" value="RadOfDeg(45)"/>
    <define name="MAX_PITCH" value="RadOfDeg(45)"/>
  </section>

  <section name="adc" prefix="ADC_CHANNEL_">
    <define name="IR1" value="ADC_0"/>
    <define name="IR2" value="ADC_1"/>
    <define name="IR_TOP" value="ADC_2"/>
    <define name="IR_NB_SAMPLES" value="16"/>
  </section>

  <section name="BAT">
```

```xml
    <define name="MILLIAMP_PER_PERCENT" value="0.86"/>
    <define name="CATASTROPHIC_BAT_LEVEL" value="9.3" unit="V"/>
</section>

<section name="MISC">
  <define name="CARROT" value="5." unit="s"/>
  <define name="NOMINAL_AIRSPEED" value="20." unit="m/s"/>
  <define name="KILL_MODE_DISTANCE" value="(1.5*MAX_DIST_FROM_HOME)"/>
  <define name="CONTROL_RATE" value="60" unit="Hz"/>
  <define name="ALT_KALMAN_ENABLED" value="TRUE"/>
  <define name="XBEE_INIT" value="\"ATPL3\rATRN1\rATTT80\r\""/>
  <define name="UNLOCKED_HOME_MODE" value="TRUE"/>
  <define name="RC_LOST_MODE" value="PPRZ_MODE_AUTO2"/>
  <define name="DEFAULT_CIRCLE_RADIUS" value="75."/>
</section>

<section name="VERTICAL CONTROL" prefix="V_CTL_">
  <define name="POWER_CTL_BAT_NOMINAL" value="11.1" unit="volt"/>
  <!-- outer loop proportional gain -->
  <define name="ALTITUDE_PGAIN" value="-0.2"/>
  <!-- outer loop saturation -->
  <define name="ALTITUDE_MAX_CLIMB" value="2."/>
  <!-- auto throttle inner loop -->
  <define name="AUTO_THROTTLE_NOMINAL_CRUISE_THROTTLE" value="0.70"/>
  <define name="AUTO_THROTTLE_MIN_CRUISE_THROTTLE" value="0.55"/>
  <define name="AUTO_THROTTLE_MAX_CRUISE_THROTTLE" value="0.80"/>
  <define name="AUTO_THROTTLE_LOITER_TRIM" value="1500"/>
  <define name="AUTO_THROTTLE_DASH_TRIM" value="-4000"/>
  <define name="AUTO_THROTTLE_CLIMB_THROTTLE_INCREMENT" value="0.15" unit="%/(m/s)"/>
  <define name="AUTO_THROTTLE_PGAIN" value="-0.01"/>
  <define name="AUTO_THROTTLE_IGAIN" value="0.1"/>
  <define name="AUTO_THROTTLE_PITCH_OF_VZ_PGAIN" value="0.05"/>
  <define name="THROTTLE_SLEW_LIMITER" value="2" unit="s"/>
</section>

<section name="HORIZONTAL CONTROL" prefix="H_CTL_">
  <define name="COURSE_PGAIN" value="-1.56"/>
  <define name="ROLL_MAX_SETPOINT" value="0.6" unit="radians"/>
  <define name="PITCH_MAX_SETPOINT" value="0.4" unit="radians"/>
  <define name="PITCH_MIN_SETPOINT" value="-0.4" unit="radians"/>
  <define name="PITCH_PGAIN" value="-15633.0"/>
  <define name="PITCH_DGAIN" value="0"/>
  <define name="ROLL_PGAIN" value="10567"/>
  <define name="ELEVATOR_OF_ROLL" value="1250"/>
</section>

<section name="AGGRESSIVE" prefix="AGR_">
  <define name="BLEND_START" value="20"/>
  <define name="BLEND_END" value="10"/>
  <define name="CLIMB_THROTTLE" value=".99"/>
  <define name="CLIMB_PITCH" value="0.4"/>
  <define name="DESCENT_THROTTLE" value="0.1"/>
  <define name="DESCENT_PITCH" value="-0.25"/>
```

```
    <define name="CLIMB_NAV_RATIO" value="0.8"/>
    <define name="DESCENT_NAV_RATIO" value="1.0"/>
  </section>

  <section name="TCAS" prefix="TCAS_">
    <define name="TAU_TA" value="10." unit="s"/>
    <define name="TAU_RA" value="6." unit="s"/>
    <define name="ALIM" value="15." unit="m"/>
    <define name="DT_MAX" value="2000" unit="ms"/>
  </section>

  <section name="FAILSAFE" prefix="FAILSAFE_">
    <define name="DELAY_WITHOUT_GPS" value="2" unit="s"/>
    <define name="DEFAULT_THROTTLE" value="0.3" unit="%"/>
    <define name="DEFAULT_ROLL" value="0.3" unit="rad"/>
    <define name="DEFAULT_PITCH" value="0.5" unit="rad"/>
  </section>

  <section name="DATALINK" prefix="DATALINK_">
    <define name="DEVICE_TYPE" value="XBEE"/>
    <define name="DEVICE_ADDRESS" value="...."/>
  </section>

  <section name="Takeoff" prefix="Takeoff_">
    <define name="Height" value="30" unit="m"/>
    <define name="Speed" value="15" unit="m/s"/>
    <define name="Distance" value="4" unit="m"/>
    <define name="MinSpeed" value="5" unit="m/s"/>
  </section>

 <makefile>
CONFIG=\"tiny_2_1_1.h\" include
$(PAPARAZZI_SRC)/conf/autopilot/tiny.makefile


FLASH_MODE=IAP


ap.CFLAGS +=  -DFBW -DAP -DCONFIG=$(CONFIG) -DLED -DTIME_LED=1
ap.srcs = sys_time.c $(SRC_ARCH)/sys_time_hw.c
$(SRC_ARCH)/armVIC.c main_fbw.c main_ap.c main.c

ap.srcs += commands.c

ap.CFLAGS += -DACTUATORS=\"servos_4017_hw.h\" -DSERVOS_4017
ap.srcs += $(SRC_ARCH)/servos_4017_hw.c actuators.c

# 72MHz
ap.CFLAGS += -DRADIO_CONTROL -DRADIO_CONTROL_TYPE=RC_FUTABA
ap.srcs += radio_control.c $(SRC_ARCH)/ppm_hw.c

# Maxstream API protocol
ap.CFLAGS += -DDOWNLINK -DUSE_UART1
-DDOWNLINK_TRANSPORT=XBeeTransport -DDOWNLINK_FBW_DEVICE=Uart1
-DDOWNLINK_AP_DEVICE=Uart1 -DXBEE_UART=Uart1 -DDATALINK=XBEE
-DUART1_BAUD=B9600
```

```
ap.srcs += downlink.c $(SRC_ARCH)/uart_hw.c datalink.c xbee.c


ap.CFLAGS += -DINTER_MCU ap.srcs += inter_mcu.c


ap.CFLAGS += -DADC -DUSE_ADC_0 -DUSE_ADC_1 -DUSE_ADC_2 -DUSE_ADC_3
ap.srcs += $(SRC_ARCH)/adc_hw.c


# basic formation
ap.CFLAGS += -DTRAFFIC_INFO
ap.srcs +=traffic_info.c


# Traffic Collision Avoidance System
ap.CFLAGS += -DTCAS
ap.srcs += tcas.c


# adding stuffs about communication with Gumstix+MNAV Haiyang
20080507


ap.CFLAGS += -DUGEAR -DXSENSDL -DUGEAR_LED=2 -DUSE_UART0
-DUGEAR_LINK=Uart0 -DUART0_BAUD=B115200


ap.srcs += osam_imu_ugear.c gps.c latlong.c


ap.CFLAGS += -DINFRARED ap.srcs += infrared.c estimator.c


ap.CFLAGS += -DNAV -DAGR_CLIMB -DLOITER_TRIM -DALT_KALMAN
-DWIND_INFO


ap.srcs += nav.c fw_h_ctl.c fw_v_ctl.c
ap.srcs += nav_line.c nav_survey_rectangle.c OSAMNav.c


# Config for SITL simulation include
$(PAPARAZZI_SRC)/conf/autopilot/sitl.makefile


sim.CFLAGS += -DCONFIG=\"tiny.h\" -DAGR_CLIMB -DLOITER_TRIM
-DALT_KALMAN -DTRAFFIC_INFO -DTCAS


sim.srcs += nav_survey_rectangle.c nav_line.c traffic_info.c
tcas.c OSAMNav.c


  </makefile>
</airframe>
```

A.3.5   Roll-Channel Fractional Order Controller Implementation

The roll-channel flight controller design part is introduced in Chapter 4. The implementation details are covered in this manual. Related files in the airborne directory are modified or added.

- *main_ap.c*: modified to replace the default flight controller with ours, called AggieController

- *ap_downlink.h*: modified to add the debug telemetry since the low-level flight control requires the real-time data for further analysis

- *osam_aggie_controller.c* and *osam_aggie_controller.h*: added to implement the low-level Aggie controllers.

The pseudo codes added in the *main_ap.c* are shown in the following.

Modifications in *main_ap.c*

```
#ifndef AGGIECONTROLLER
        h_ctl_attitude_loop();
        v_ctl_throttle_slew();
#else
        if (pprz_mode != PPRZ_MODE_AUTO1) {
                aggie_h_ctl_setpoint();
                aggie_h_ctl_roll_loop();
                aggie_h_ctl_pitch_loop();
                aggie_ctl_throttle_loop();
                . . .

        }
#endif /* AGGIECONTROLLER*/
```

The main parts for the implementation of the proposed FOPI controller is achieved in the *osam_aggie_controller.c* file, shown in the following.

```
/* $Id: osam_aggie_controller.c 2710 2008-09-12 22:11:35Z osam $
 * Copyright (C) 2009-2019  Haiyang Chao
 */
/** \file osam_aggie_controller.c
 *  \brief header file for Aggie flight controller
 * This file is first generated by Haiyang Chao on 20090929.
 */

/*aggie_outer_roll_loop function is to do a pid control of roll
loop from a outter loop*/

void aggie_outer_roll_loop_pid(void){
   float h_ctl_roll_outer_pgain = ir_correction_left;
   float h_ctl_roll_outer_igain = ir_correction_right;
   static float i = 0.0; /*Ki controller output*/

   BoundAbs(aggie_roll_outer_setpoint, h_ctl_roll_max_setpoint);
   float p = h_ctl_roll_outer_pgain*(estimator_phi - aggie_roll_outer_setpoint);
   float v = p + i;
   aggie_roll_setpoint = SAT2(v,PI/4*ir_correction_down);
```

```
  i = i+h_ctl_roll_outer_pgain*h_ctl_roll_outer_igain/FREQ*
  (estimator_phi-aggie_roll_outer_setpoint)+
  2*h_ctl_roll_outer_igain/FREQ*(aggie_roll_setpoint-v);

  /*The following is just for debugging use
  aggie_float1 = 111.1;
  aggie_float2 = i;
  aggie_float3 = v;
  aggie_float4 = aggie_roll_setpoint;*/
}


/*aggie_outer_roll_loop_foc function is to do a pid control of
roll loop from a outter loop*/

void aggie_outer_roll_loop_foc(void){
  float h_ctl_roll_outer_pgain = ir_correction_left;
  float h_ctl_roll_outer_igain = ir_correction_right;
  static float err_windup = 0.0;
  static float i = 0.0;/*variable for integrator output*/

  BoundAbs(aggie_roll_outer_setpoint, h_ctl_roll_max_setpoint);
  float p = h_ctl_roll_outer_pgain*(estimator_phi - aggie_roll_outer_setpoint);
  u_foc2 = u_foc1; /*update u(t-2)*/
  u_foc1 = u_foc0; /*update u(t-1)*/
  u_foc0 = p*h_ctl_roll_outer_igain+2*h_ctl_roll_outer_igain*err_windup;
  y_foc2 = y_foc1; /*update y(t-2)*/
  y_foc1 = y_foc0; /*update y(t-1)*/

  float foc = u_foc0*NUM2+u_foc1*NUM1+u_foc2*NUM0-y_foc1*DEN1-y_foc2*DEN0;
  y_foc0 = foc;
  i = i + y_foc1/FREQ;
  float v = p + i;
  aggie_roll_setpoint = SAT2(v,PI/4*ir_correction_down);
  err_windup = aggie_roll_setpoint - v;

/* for debug use only*/
  aggie_float1 = foc; //for debug 20091009
  aggie_float2 = i; //for debug 20091009
  aggie_float3 = u_foc0; //for debug 20091009
  aggie_float4 = v; //for debug 20091009
} /* function aggie_outer_roll_loop_foc stopped here*/

void aggie_h_ctl_roll_loop( void ) {
  float err = estimator_phi - aggie_roll_setpoint;
  float cmd = h_ctl_roll_pgain * err;
  h_ctl_aileron_setpoint = TRIM_PPRZ(cmd);
}

void aggie_h_ctl_pitch_loop( void ) {
  static float last_err;
  float err = estimator_theta - h_ctl_pitch_setpoint;
  float d_err = err - last_err;
  last_err = err;
```

```
    float cmd = h_ctl_pitch_pgain * (err + h_ctl_pitch_dgain * d_err);
    h_ctl_elevator_setpoint = TRIM_PPRZ(cmd);
}

void aggie_ctl_throttle_loop( void ) {
  float cmd = 6720.0; /*70% 20091002*/
  v_ctl_throttle_slewed = TRIM_UPPRZ(cmd);
}
```

A.4 DIFFMAS2D CODE MODIFICATION GUIDE

DiffMas2D [7] is a diffusion control simulation platform written in MAT-LAB/Simulink. The basic idea of DiffMas2D is to use a sensor and actuator network for the sensing and actuation of a diffusion process modeled by the following equation:

$$\frac{\partial \rho}{\partial t} = k \left(\frac{\partial^2 \rho}{\partial x^2} + \frac{\partial^2 \rho}{\partial y^2} \right) + f_d(x, y, t) + f_c(\tilde{\rho}, x, y, t), \tag{A.1}$$

where k is a positive constant representing the diffusing rate, $f_d(x, y, t)$ shows the pollution source, $\tilde{\rho}$ is the sensor measurements, $f_c(\tilde{\rho}, x, y, t)$ is the control input applied to neutralize pollutants.

The major functions of the DiffMas2D software include

- solving the PDE model (A.1)
- providing sensor readings for any position within the domain
- applying the feedback control law on how much to spray ($f_c(\tilde{\rho}, x, y, t)$)
- planning the path for the actuators.

A.4.1 Files Description

DiffMas2D comprises one Simulink model and a series of preprocessing and post-processing M files. The main files are briefly introduced in the following:

- *simstart.m*: the file to start the simulation
- *initialization.m*: the file to set up the initial parameters such as the diffusion speed, sensor and actuator numbers, and initial positions
- *init_check.m*: the file to test if there is any conflictions between the settings and the Simulink model
- *pre_process.m*: the file for the boundary settings
- *diffu_ctrl_sim.mdl*: the main model file for the real-time simulation
- *post_process.m*: the file to depict all the saved data in pictures.

diffu_ctrl_sim.mdl is the central file of the DiffMas2D simulation platform, shown in Fig. A.4 and the key subfunctions are defined with M files such as the

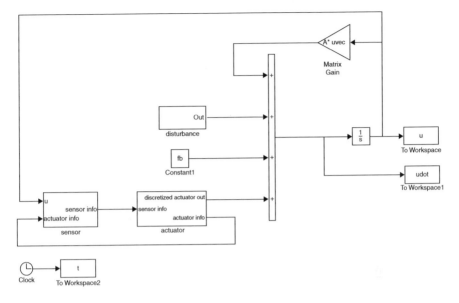

Figure A.4 DiffMAS2D Simulink model.

actuator path planning and the spraying control. Several important M functions are introduced in the following:

- *controller.m*: the function for neutralizing control
- *actrl.m*: the function for the actuator position control
- *ades.m*: the function to calculate the desired position of each actuator for monitoring purposes
- *distout.m*: the function to define the diffusion source.

A.4.2 Diffusion Animation Generation

DiffMAS2D supports the generation of the diffusion animation after running the simulation for a preset time period. All the simulated data are saved in the workspace and *post_process.m* can convert the data into jpg pictures. The user can then use some softwares to convert multiple jpg pictures into an animation gif file.

A.4.3 Implementation of CVT-Consensus Algorithm

The CVT-Consensus algorithm is mostly implemented in *actrl.m* and *ades.m*. The *actrl.m* file is shown in the following.

```
function out = actrl(in)
% actrl: actuator position control. Output the acceleration of each actuator.
```

```
%    The actuator is modeled as two double integrators:
%                    $\ddot x=f_x(t)$,
%                    $\ddot y=f_y(t)$.
%    $f_x(t)$ and $f_y(t)$ are defined in this file.

%  keyboard
in = in(:);


NS = in(1); % number of sensors
NA = in(2); % number of actuators
apos = reshape(in(3:2*NA+2), NA, 2);


% actuator velocity matrix. [avel(i,1),avel(i,2)] is the current velocity
% of the ith actuator
avel = reshape(in(2*NA+3:4*NA+2), NA, 2);


% current sensor infomation.
% [sinfo(i,1),sinfo(i,2)] is the current position of the ith sensor
% [sinfo(i,3),sinfo(i,4)] is the current velocity of the ith sensor
% sinfo(i,5) is the sensed data of the ith sensor
sinfo = reshape(in(4*NA+3:4*NA+5*NS+2), NS, 5);


% current time
t = in(end);


%%%%%%%%%%%%%%%%%%%%%%%%%%%%%%%%%%%%%%%%%%%%%%%%%%%%%%%%%%%%%%%%%%%%%%%%%%%%%%
% Code below this line should be written by the user to achieve the desired  %
% actuator movement. The final output is a vector in the format of          %
% [fx_1,fx_2,...,fx_NA,fy_1,fy_2,...,fy_NA]'                                 %
%%%%%%%%%%%%%%%%%%%%%%%%%%%%%%%%%%%%%%%%%%%%%%%%%%%%%%%%%%%%%%%%%%%%%%%%%%%%%%
global p_out sum_s controlTime; controlTime=controlTime+1;
if controlTime==100 %300
    controlTime=0;
    %arrays to record the points that is most close to i's actuator
    p=zeros(NA, 300);
    %array to indicate the number of points for each actuator
    p_index=zeros(NA,1);
    %find the Vi for each actuator
    p_out=zeros(NA,2);
    sum_s=zeros(NA,1);
    %design acceleration for each actuator
     k1=3;   k2=1;

    % Voronoi Diagram computing (Nearest actuator strategy)
    for i=1:NS
        min=( ( apos(1,1)-sinfo(i,1) )^2+( apos(1,2)-sinfo(i,2) )^2 );
        mini=1;
        for j=2:NA
            temp=( ( apos(j,1)-sinfo(i,1) )^2+( apos(j,2)-sinfo(i,2) )^2 );
            if temp < min
                min=temp;
                mini=j;
            end
```

```
        end
        if(min < 0.04)  % added this condition 20060923 for increasing PrayK
            p_index(mini)=p_index(mini)+1;
            p( mini,p_index(mini) )=i;
        end
    end
    des=zeros(NA,2);

    for i=1:NA
        sumx=0;
        sumy=0;
        sumd=0;
        for j=1:p_index(i)
            % added 20060903 to delete positions that sensor reading too small
                sumx=sumx+sinfo(p(i,j),5)*sinfo( p(i,j),1);
                sumy=sumy+sinfo(p(i,j),5)*sinfo( p(i,j),2);
                sumd=sumd+sinfo(p(i,j),5);
        end

        if(sumd == 0)
            des(i,1)=apos(i,1);
            des(i,2)=apos(i,2);
            sum_s(i)=0;
        else
            des(i,1)=sumx/sumd;
            des(i,2)=sumy/sumd;
            sum_s(i)=sumd/p_index(i);
        end
    end
    if(t<2)
        Lmatrix3 = [-2 0 -1 0;0 -2 -1 0;0 0 -3 0;0 0 -1 -2];
        p_out = Lmatrix3*(apos-des);
    elseif ((t > 2) && (sum(sum_s/sum_s(3)) < 2))
        Lmatrix1 = [-1 0 1 0;0 -1 1 0;0 0 -1 0;0 0 1 -1];
        Lmatrix2 = [0 0 0 0;0 0 0 0;0 0 -1 0;0 0 0 0 ];
        p_out = Lmatrix1*apos-Lmatrix2*des;
    else
        Lmatrix0 = (-1)*k1*eye(NA,NA);
        p_out = Lmatrix0*(apos-des);
    end
end

if(t > 0.2)
    fgorki = 1;
end out = reshape(p_out, 2*NA,1);
```

REFERENCES

1. u-blox company. u-blox GPS protocol, 2010. http://www.u-blox.com.
2. Xbow Technology. MNAV IMU specifications, 2009. http://www.xbow.com.

3. C. L. Olson. Microgear open source project, 2008. http://cvs.flightgear.org/ curt/UAS/ MicroGear1.

4. Microstrain Inc. Gx2 IMU specifications, 2008. http://www.mirostrain.com.

5. Xsens Company. Xsens-Mtig IMU specifications, 2008. http://www.xsens.com.

6. Open Source Paparazzi UAV Project, 2008. http://www.recherche.enac.fr/paparazzi/.

7. J. Liang and Y. Q. Chen. Diff-MAS2D (version 0.9) user's manual: a simulation platform for controlling distributed parameter systems (diffusion) with networked movable actuators and sensors (MAS) in 2D domain. Technical Report USU-CSOIS-TR-04-03, CSOIS, Utah State University, 2004.

Topic Index

Remote Sensing and Actuation Using Unmanned Vehicles, First Edition. Haiyang Chao and YangQuan Chen.
© 2012 by The Institute of Electrical and Electronics Engineers, Inc.
Published 2012 by John Wiley & Sons, Inc.

IEEE PRESS SERIES ON SYSTEMS SCIENCE AND ENGINEERING

Editor:
MengChu Zhou, *New Jersey Institute of Technology and Tongji University*

Co-Editors:
Han-Xiong Li, *City University of Hong-Kong*
Margot Weijnen, *Delft University of Technology*

The focus of this series is to introduce the advances in theory and applications of systems science and engineering to industrial practitioners, researchers, and students. This series seeks to foster system-of-systems multidisciplinary theory and tools to satisfy the needs of the industrial and academic areas to model, analyze, design, optimize and operate increasingly complex man-made systems ranging from control systems, computer systems, discrete event systems, information systems, networked systems, production systems, robotic systems, service systems, and transportation systems to Internet, sensor networks, smart grid, social network, sustainable infrastructure, and systems biology.

1. *Reinforcement and Systemic Machine Learning for Decision Making*
   Parag Kulkarni
2. *Remote Sensing and Actuation Using Unmanned Vehicles*
   Haiyang Chao, YangQuan Chen
3. *Hybrid Control and Motion Planning of Dynamical Legged Locomotion*
   Nasser Sadati, Guy A. Dumont, Kaveh Akbari Hamed, and William A. Gruver

Forthcoming Titles:

Operator-based Nonlinear Control Systems Design and Applications
Mingcong Deng

Contemporary Issues in Systems Science and Engineering
Mengchu Zhou, Han-Xiong Li and Margot Weijnen

Design of Business and Scientific Workflows: A Web Service-Oriented Approach
Mengchu Zhou and Wei Tan

Printed and bound by CPI Group (UK) Ltd, Croydon, CR0 4YY